Moral Cosmology

Moral Cosmology

On Being in the World
Fully and Well

Albert Borgmann

LEXINGTON BOOKS
Lanham • Boulder • New York • London

Published by Lexington Books
An imprint of The Rowman & Littlefield Publishing Group, Inc.
4501 Forbes Boulevard, Suite 200, Lanham, Maryland 20706
www.rowman.com

86-90 Paul Street, London EC2A 4NE

British Library Cataloguing in Publication Information Available

Library of Congress Cataloging-in-Publication Data

Names: Borgmann, Albert, author.
Title: Moral cosmology : on being in the world fully and well / Albert Borgmann.
Description: Lanham : Lexington Books, [2023]. | Includes bibliographical references.
   | Summary: "This book argues for a unified worldview of moral cosmology that will
   allow us to be truly at home in the universe, a view that was disrupted by the European
   Enlightenment. The author contends that a basic understanding of quantum physics
   and relative theory offers the widest possible background for the renewal of a moral
   cosmology"— Provided by publisher.
Identifiers: LCCN 2023034653 (print) | LCCN 2023034654 (ebook) | ISBN
   9781666900460 (cloth) | ISBN 9781666900477 (ebook)
Subjects: LCSH: Quantum cosmology—History. | Physics—Moral and ethical aspects. |
   Mathematics—Moral and ethical aspects.
Classification: LCC QB991.Q36 .B67 2023 (print) | LCC QB991.Q36 (ebook) | DDC
   170—dc23/eng/20231214
LC record available at https://lccn.loc.gov/2023034653
LC ebook record available at https://lccn.loc.gov/2023034654

For My Beloved

Contents

Acknowledgments

I'm indebted to my daughter Caitlin Borgmann; my friends and colleagues, Pat Burke, Daniel Kemmis, Dan Spencer and Jason Wiener; my research assistant, Max Barton; Jana Hodges-Kluck; and an anonymous reviewer for helping the manuscript become a book.

Introduction

When you wake up in the morning, you are face-to-face with the chores of the day. Your tasks and worries take you through the day. In the evening you relax and go to bed, and in the morning, you are once more up against the day's demands. So how could you possibly be concerned with the universe, far less with its moral force? The first step toward an answer is the realization that your daily conduct proceeds against an encompassing if implicit background. Its force occasionally becomes explicit as it does in the Film *On the Beach* (1957) where certain death from a radioactive cloud that has been caused by nuclear war is descending on Australia from the north.[1]

At the personal level too, you are sometimes overtaken by definite background knowledge through poignant events, joyful ones like the birth of a child or a promotion at your job, sad or crushing events like the death of a beloved or the loss of your home. Such an emergence of background knowledge is limited, however, to you, your friends, and at most to some public region. The COVID-19 pandemic has taught us otherwise. It's a universal and worrisome and sometimes terrifying and crushing burden. But by now we can be confident it will pass as a major force. What lasting effect will it have? It depends on whether we experience the pandemic as a nightmare we want to forget as soon as possible or as a crucible that has brought out the best in us.

But even in the best case, the pandemic does not by itself open up cosmic awareness. So why worry about it? We can learn the importance of understanding the cosmos from our primeval culture, the hunting and gathering culture that was the human way of life for hundreds of centuries. We can tell from stories and archeological remainders that cosmic awareness was then powerful and well-informed. To know the course of the Sun and the planets through the year was a matter of course. The divinity and the moral rules of the heavenly bodies were taken for granted.

The European Enlightenment divided moral cosmology into physics and ethics. Physics tells us what is, ethics what we ought to do. Neither of the two disciplines has jurisdiction over the other.[2] But comprehending physics has become hard, and understanding ethics has become shifting and uncertain. Truly to inhabit the world demands that we search for one world so that we learn to be at home in the world once again. A necessary condition for so being in the world is a rudimentary knowledge of physics, of the very small through quantum physics and of the very large through relativity theory. To teach such knowledge is a pedagogical task that we're just beginning to take on.

What about the sufficient condition of fully being in the world? The great popularizer of mathematics and physics Martin Gardner and the Nobel Prize–winning physicist Steven Weinberg tell us that the real world is devoid of ethics and pointless.[3] But that kind of the world's reality is artificially cleansed of its moral force. That leaves us with the task of discovering a cosmic moral center that is compatible with physics to be sure, but also enables us to be in the world fully and well.

To clear the air for what follows in these pages, let me distinguish between moral cosmology and cosmic ethics. Several authors have been concerned with cosmic ethics, and naturally they don't all agree on what cosmic ethics means. The term has been used for grand proposals as well as a synonym for contemporary global ethics; but unlike moral cosmology, cosmic ethics entails neither a concern regarding the separation of ethics from physics nor the aspiration of undoing the division.[4] You can't, they say, get an *ought* out of an *is*. Immanuel Kant has given this division its paradigmatic imprint by sharply dividing the function of pure reason into two parts, *theoretical reason*, entrusted with the investigation of nature, and *practical reason*, empowered to formulate the categorical imperative.[5] Jacques Maritain in his great 1964 book *Moral Philosophy* discusses exceptions to this rule, authors who have attempted—unsuccessfully in Maritain's view—to unify physics and ethics into a cosmic ethics. Maritain chiefly discusses the work of Dewey and Bergson.[6] There are, to be sure, affinities, however distant, between cosmic ethics and moral cosmology.

NOTES

1. The film is based on the novel by Nevil Shute, *On the Beach* (Heinemann, 1957).

2. Steven Jay Gould, *Rocks of Ages: Science and Religion in the Fullness of Life* (New York: Ballantine Books, 2011).

3. Martin Gardner, *Relativity for the Million* (New York: Pocket Books, 1965 [1962]), pp. vii–viii. Steven Weinberg, *The First Three Minutes*, 2nd ed. (New York: Basic Books, 1993 [1977]), p. 154.

4. William Cave Thomas (1820–1906; the year of Thomas's death seems uncertain, but 1906 seems most likely), *Ethics or the Mathematical Theory of Evolution Showing the Full Importance of the Doctrine of the Mean, and Containing the Principia of the Science of Proportion* (London: Smith, Elder, & Co., 1896). In the year of its publication, E.E.C. Jones remarked that Thomas's proposal was "not only unproven, but also unprovable" in *International Journal of Ethics*, volume 7 (1896), p. 514. As for the use of "Cosmic Ethics" by other authors, see Gordon Arthur, "Religion and Values: Cosmic or Universal Ethics?" *Journal of Space Philoso-*

phy, volume 3, no. 2 (Fall 2014), pp. 23–30; Kelly Smith, "Cosmic Ethics: A Philosophical Primer," https://www.oocities.org/dwmyatt/ethics-life2.html, consulted August 30, 2021. *Theological Implications of Astrobiology,* ed. C. Bertka, N. Roth, and M. Shindell (Washington, AAAS); "The New Cosmic Ethics: Morality of the Future," in "What Are Ethics" (2009), consulted August 30, 2021.

    5. Immanuel Kant, *Kritik der reinen Vernunft,* ed. Raymund Schmidt (Hamburg: Felix Meiner, 1956 [first edition of the *Kritik,* 1781]; *Kritik der praktischen Vernunft,* ed. Karl Vorländer (Hamburg: Felix Meiner, 1963 [first edition of this *Kritik,* 1787].

    6. Jacques Maritain, *Moral Philosophy* (London: Geofrey Bles, 1964), pp. 396–447.

1

The Rise and Fall of Moral Cosmology

THE RISE OF MORAL COSMOLOGY

The evolution of human consciousness must have been like a sunrise—the gradual emergence of reality from darkness and dimness into the sharp articulation of broad daylight. Or, correspondingly, it was like a slow awakening, things coming into focus little by little. To be sure, while this was happening over the span of a million years or so, our ancestors were fully alive to their environment and skilled in coping with it. Human consciousness dawned while hominins were wholly awake to their world. But the rise of conscious awareness widened our view of the world, and as soon as consciousness had come into the open, humans must have been struck by the power of the Sun, the waxing and waning of the Moon, and the depth of the starry sky.

The night, presumably, was the time when humans began to wonder about the extent and the order of the world. The day is busy and full of things and tasks. The course and the brilliance of the Sun determine things on earth but leave the sky blue and blank. A cloudless night, to the contrary, opens up endless expanses and naturally provokes questions. Is there an order to this? Has it always been like this? How far does it go? Where do we fit into all of this? The first answers to these questions were given a hundred millennia ago or more and they have been lost forever. But a few tangible records of that time have survived, bones marked with parallel incisions, and the oldest (44,000 to 43,000 years BCE), the Lebombo bone so called, may in fact be a testament to cosmological awareness.[1]

As soon as humans told and handed down stories, there have been cosmologies. They have come down to us either as written records or as accounts from the oral traditions of hunting and gathering tribes. Perhaps the most widely known ancient cosmology is the biblical story of Genesis. It explains the moral significance of the universe by telling us how it came to be. The world came into being through a divine creator. It is well-ordered and good. Humans are the crown of creation, and the world is entrusted to them. The moral cosmology of Genesis then, inspires gratitude to the creator, confidence in the order and goodness of the world, and determination in cultivating nature.

In Genesis, God set the Sun in the firmament of the heavens to rule the day and the seasons. For the Hebrews the Sun was evidence of God's power, wisdom, and benevolence. For the ancient Greeks and the Blackfeet of North America (as for many other premodern cultures), the Sun itself was divine, the god Helios for the Greeks, Sun Chief for the Blackfeet.

There was mythic commerce among ancient cultures. Local worship of the Sun arose in different places and with different names for the divinity. Local worship was exported and modified by imports. The diversity of local cults was often simplified and systematized. The worship of the Sun in ancient Greece was pushed to the periphery through the ascendancy of the Olympic gods. Still, we can tell from accounts in *The Odyssey*, from the poetry of Ovid, and from incidental remarks of other authors that the ancients marveled at the tireless and powerful movement of the god and thought of themselves as inescapably watched by the all-seeing Helios.

For the Blackfeet, Sun Chief was the ruler of the world, given to righteous anger, but ultimately forgiving and benevolent not unlike God in the Hebrew scriptures.[2] The focal point for the entire Blackfoot Nation was the annual Sun Dance, an occasion for strenuous devotion, communal celebration, and social pleasure. The Sun Dance was a time to pay one's debt for divine favors, painfully so for the protagonist in James Welch's story, to revive awareness of the spirit world, and to reinvigorate bonds of friendship.[3]

Whether Hebrew, Greek, or Native American, moral cosmologies were eminently local, centered on a place like the temple in Jerusalem, Mount Olympus, or the Four Persons Butte near the Milk River in northern Montana. Those places were animated by divinity, by the transcendent Yahve or the daily and reassuring presence of Helios or Sun Chief. And the cohesion and continuity of cosmologies were secured by the stories that were handed from generation to generation. Just as the ancient cosmologies did not distinguish between ethics and physics, neither did they divide space from time or time from order.[4]

TODAY'S LACK OF MORAL COSMOLOGY

If you've been born here or have been naturalized, you're a citizen of this country. But to be more than a citizen in name only and merely by virtue of a document, you should know this country. You should know who John Winthrop was, what the Declaration of Independence says, what the GOP is, and where Yellowstone Park is located. We are not only citizens of this country; however, we are also citizens of the world. We should be able to locate China on an unmarked map, we should know why ancient Greece is important, we should be able to name the founder of Islam, we should have read the charter of the United Nations.

Everyone should have the benefit of this kind of general education. There is in fact a program of "general education" in most colleges and universities though all too often it lacks substance.[5] Still, there is something like an educated class in the United States whose members can claim to know the crucial elements and dimensions of American and global geography, history, economics, politics, and culture. The national conversation that's carried on in, say, *The New York Times*, *The Wall Street Journal*, and *The New Yorker* is evidence of the depth and extent of national and global citizenship. There are countless media outlets, educational institutions, think tanks, conferences, lectures, and conversations in bars, on planes, and around the dinner table that testify to a remarkable and encouraging fact—there must be millions of people who have earnestly and admirably appropriated their closer and wider world. A careful observer can get a feel for the typical extent and quality of this sort of local and cosmopolitan citizenship. Remarkably, however, even this serious and high-minded kind of world appropriation rarely extends more than a few miles above the surface of the planet. It extends no farther out into space than to the atmosphere that is now causing the earth to warm and to the satellites that connect and keep track of what matters on earth.

People by and large have no cosmology, i.e., they have no conception of the structure of the universe and of the place they occupy in it. "The place of humans in the cosmos" refers first of all to our coordinates in cosmic time and space. But when we inquire into the place of one thing in the context of another, e.g., of the place of physics in general education, we are asking about the significance of something in a larger context. The ultimate and decisive significance of a person or a thing is by common agreement their moral significance. Hence what is perhaps most remarkable today is the common lack of a moral cosmology even among the astrophysicists who know all there is to know about the physical place of the earth in the universe. The contemporary absence of a moral cosmology contrasts with its presence in most all premodern cultures.[6] They had a story and a conception of the world as a whole, how it came to be, what its crucial forces and dimensions were, and how the world at large informed the world down here and instructed the conduct of humans. By our lights, of course, these primeval cultures contained limited and mistaken conceptions of the cosmos. Yet these worldviews extended beyond the atmosphere and comprehended everything that was within reach.

A plausible story can be told why there is no common moral cosmology today. The price of a moral cosmology is error, one might argue. It's simply wrong to say that on the fourth day of creation God set the Sun in the firmament of heaven or that Sun Chief oversees everything on earth and cares for the welfare of humans. The Sun is in fact a medium-size star that is a remnant of the Big Bang that occurred 13.7 billion years ago. Physics is one thing, ethics is another, and cosmology is just physics. Moreover, the grand book of the universe, as Galileo remarked,

> is written in the language of mathematics, and its characters are triangles, circles, and other geometrical figures without which it is humanly impossible to understand a single word of it; without these, one is wandering about in a dark labyrinth.[7]

Equipped with high school geometry, we might be able to follow Galileo's geo-metrical account of the cosmos. But the total mathematics of quantum theory and relativity theory is difficult. The mathematics of the theory that attempts to unify those two theories, string theory, is forbidding and understood by just a handful of experts. When string theory first emerged as a possible unification of the presently incompatible theories of quantum mechanics and general relativity, it "was so daunt-ing a task that all but the most courageous physicists recoiled at the challenge" as Brian Greene put it, who then went on to say: "A number of us consistently worked deep into the night to try to master the vast areas of theoretical physics and abstract mathematics that are required to understand string theory."[8] So what is a layperson with a day job supposed to do?

But aren't there shelves of books by first-rate physicists that explain current as-trophysics without a single equation? Yes, you bet, and all of us lay folks are deeply indebted to them. But even those accounts are conceptually tough because they are at odds with the "naive physics" (also called "folk physics") we have evolved with. Folk physics is a deeply embedded set of concepts that help us grasp what the imme-diate world is like and how it operates. One-year-olds understand that objects hold together and are subject to gravity and causality. Children spontaneously believe that events happen for a reason.

Our traditional adult view of the world grows out of our naive physics. As each of us knows from experience, we are natural geocentrists. The sun appears to rise and to set no matter how often we have been told that in reality our spot on earth turns toward and away from the sun. What's worse, we are inveterate flat-earthers. The up-and-down dimension is essential to the ways we appropriate our world; that's evident in the count-less metaphors that provide orientation for our speaking—good things rise; falling things are bad; what is in front of you is public, what's in back is private, etc.[9] And here too, being told that we in North America are "down under" for people in southern Chile does not dispel our sense that Chileans are really down there. We know some things are fast and others faster yet. That there should be an absolute limit to how fast anything can move does not seem natural. When the clock strikes twelve, not everyone in the universe will say it's now "noon," but it does seem natural to think that it's the very same moment for all of them, no matter whether or how they mark it.

Traditional cosmology is centered on the family and from there extended to band or village, to tribe, and finally to the nation. Native Americans saw it that way as James Welch has taught us—the family of Rides-at-the-Door within the band of the Lone Eaters, within the tribe conquered or raided, and beyond that the unknown, too vast, too wild, or too insignificant to bother with. Vertically, the native and naive world was positioned midway between the gods and the demons, between Sun Chief and the Under-Water-People, between heaven and hell. People knew everything there was to know at the time, from the smallest to the largest. Theirs was a world wherein the moral and material dimensions were the same or, to put it learnedly, where there was no difference between physics and ethics, and it came with a world-view that we today can neither embrace nor give up.

THE EMERGING DISTINCTION BETWEEN
PHYSICS AND ETHICS—ARISTOTLE

In tracing the dissolution of moral cosmology, I am not concerned to give anything like a comprehensive account. What's needed here is a sketch of how moral cosmology gradually divided into physics and ethics. The first crack in the traditional moral cosmologies opened up at the turn from the seventh to the sixth century BCE in the colonies of ancient Greece. Curiosity displaced piety. Not that traditional cultures were incurious. Hunters and gatherers had to be inquisitive about the places and seasons of tubers and berries and about the wiles and ways of animals. Curiosity about plants and materials gave rise to agriculture and to large settlements thousands of years before Thales (625–547 BCE) initiated philosophy in Miletus.[10]

Traditional curiosity was suffused, however, with reverence for the all-pervasive reign of divine powers. Thales stepped back from enthrallment and proposed that the world was ruled by laws of nature rather than by powers of divinity. The particulars of Thales's theory, as far as we know it, seem impossibly simple to us. All of reality, Thales taught, consisted of evaporations, condensations, and solidifications of water. The point is that Thales's world consisted essentially of one kind of substance in a definite number of states. That was it—a physics without an ethics.

We can think of Thales as planting a seed that germinated in ancient Greece and then took three thousand years to grow into a tree and attain its characteristic shape. Physics and ethics became distinguishable branches, but at first were strongly and equally rooted in one organism. The perfect instance is Aristotle's (384–322 BCE) work. Andronicus of Rhodes, who in the first century BCE edited the neglected and rediscovered writings of Aristotle, had no trouble titling sets of lecture notes as "Ethics," another set as "Physics," and yet another set as "On the Heavens." These lecture notes presumably accumulated over the years from Aristotle's own hand and from those of his students. Thus, they reflect inconsistencies, developments, and revisions. Even so, they represent a remarkably unified vision of the world.

Aristotle discussed the possibility that the earth is a planet rather than the center of the universe, and half a century after his death Aristarchus of Samos (fl. ca 280 BCE) worked out a heliocentric system. But their failure to observe a parallax relative to the fixed stars (from planet Earth the configuration of the fixed stars should look different from different points of the Earth's putative orbit around the sun) among other reasons persuaded most ancient cosmologists that the Earth was immobile and the center of the universe.

Were the ancients justified in their rejection of heliocentrism? Why was there no parallax? The answer to the question haunts scientists and philosophers and on occasion our everyday predicaments too. The Greek cosmologists were helplessly in the grip of a plausible but false assumption. There should be a parallax observable by the naked eye if the fixed stars were distant, but not unimaginably distant. So they are, however (something Aristarchus appears to have realized).

This kind of helplessness is worse when you suspect that you are making a faulty assumption, but try as you may, you can't put your finger on it. You come to the end of a repair. All you need to do is attach the shiny knob you unscrewed an hour ago. Where is it? You scour your workbench from end to end. No luck. It must have fallen on the floor. You explore the vicinity of the workbench and then every possible nook and cranny of the workroom. You move things, lift things, sweep things. Nothing. What's the wrong assumption you're making? Then one fine winter afternoon something is glittering in the bird feeder. It's your knob. How did it get there? It was neither on your bench nor on the floor somewhere. It fell into a bag of sunflower seeds under the bench, the bag sticking out just enough to catch the falling knob silently.

Locating an unwarranted assumption has become an intriguing move because Immanuel Kant made ostensibly triumphant use of one in the *Critique of Pure Reason*, thus proposing to resolve long-standing and seemingly unresolvable disputes about whether the world is finite or infinite, granular or smooth, wholly physical or also and irreducibly ethical, and based on necessity or on contingency.[11] Though today few philosophers agree with Kant's particulars in his resolution of these four antinomies, the formal schema is powerful and attractive. Are the incompatibility of relativity and quantum theory, the disjunction of physics and ethics, and the distinction between lawfulness and contingency simply the reflections of unwitting and mistaken assumptions?

The charm of Aristotle's cosmology lies partly in a harmony that's untroubled by such doubts. To be sure, Aristotle carefully discussed all rival views he was aware of, but satisfied himself that he had a reasonable reply to each of them. He concluded that the Earth was round and resting at the center of the universe. It was surrounded by nested translucent shells, the "spheres," with one planet attached to each sphere. The spheres revolved about Earth; the shell closest to Earth carried the Moon, the outermost shell carried the fixed stars.

The spheres, being translucent, were invisible. Only the revolving planets and stars could be seen. To account for the fact that from an earthly point of view the planets do not smoothly circle about us, but at a certain point reverse course for a short while and then continue on their clockwise path (a phenomenon straightforwardly explained in a heliocentric system), Aristotle had to attach the planets to more than one sphere. These complications aside, the picture of nested spheres, cradling the earth at their center, was reassuring and ruled the imagination of the Occident for two thousand years. Aristotle materialized the cosmic framework by way of the traditional four elements—earth, water, air, and fire for the world within the lunar sphere, and with a fifth kind of matter, the quintessence, or ether, for the planets (the sun being one of them).

The design of Aristotle's cosmos was orderly from top to bottom and across the realms of motion (kinematics), animation (psychology), and perfection (ethics). Things on Earth and below the moon had a natural place according to the element they consisted of, and if displaced, they would naturally seek their place in the cosmic order. Earthy things would fall, fiery things would rise. Sideways motion was unnatural and required force. It would cease when the force was spent. Rectilinear motion betrayed imperfection, the need to reach something that was lacking. Circular motion was at one with itself and perfect. The planets, moving in perfect circles, were superior to sublunar things, and they were superior also in their matter. Ether was the finest and noblest of the elements.[12]

Anima is the Latin translation of Greek *psyche*, what we call soul. Aristotle's world was animated.[13] He thought of the soul as the life-giving principle, the energy that informs matter. Here again we find an order that agrees with the cosmic structure and the hierarchy of elements. It's the order of mineral, vegetable, animal, and human, and it extends outward to the planets that Aristotle considered divine spirits. The order of animating forms is one of increasingly vigorous world appropriation. What informs rocks is relatively inert. Being made of the element earth it occupies or seeks its place at the center. That's it. Plants have a more powerful form. It is life-giving, a soul in Aristotle's psychology. Plants appropriate minerals for their nutrition and growth and seek their perfection in the unfolding of their characteristic shape. Animal souls too possess nutrition and growth, but they also take in the world through perception and desires.

In addition to their vegetative and animal soul, humans have a rational soul.[14] Latin *rationalis* is the translation of Greek *logike*, the adjective derived from logos. Although that word makes us think chiefly of logic and logic of reasoning and reasoning of calculating, the helpful sense of the word Aristotle used is the oldest meaning of the verb that logos is derived from—*legein*, to gather. The distinctive faculty of humans is their ability to appropriate and gather the world entire, at least potentially. Hence Aristotle's celebrated words: "The soul is somehow everything."[15] Still, humans are tied to their vegetable and animal natures as well and to the brute matter of earth. Not so the planets whose powerful intelligence is materialized in the fine and pliable matter of ether. There is only one kind of still greater perfection possible, a being that is entirely unencumbered by matter, a pure form and intelligence. Such is the divinity that resides beyond the outermost sphere of the fixed stars, the ultimate source of meaning and motion that in its perfection doesn't need to go anywhere—the unmoved prime mover, the perfect, divine being.[16]

Sublunar things need to achieve their perfection, and perfection is the complete unfolding of form in matter. For humans particularly there are standards for the pursuit of perfection—the virtues. For the vegetable and animal components of the human soul there are the moral virtues of courage, justice, and friendship. The norms of excellence for the rational soul are the intellectual virtues, wisdom foremost among them. The best kind of life is devoted to wisdom, and to be wise is to comprehend the cosmos, the eternal order and beauty of the world. Such a life comes closest to the serenity and self-sufficiency of the gods.

Aristotle's cosmology dominated Western civilization for two thousand years. What the Middle Ages added to Aristotle's vision was creation. One might think that the Judeo-Christian and Islamic universe that has a beginning would be fundamentally different from Aristotle's eternal cosmos. But Thomas of Aquino (1225–1274), who revered Aristotle and was unwilling to disagree with him, thought of creation as an addition to Aristotle's view rather than an inconsistency with it. Like Augustine (354–430 CE) before him, Thomas distinguished between timeless and limitless eternity. God's eternity is timeless, and theoretically God might have created either an ever-existing or a temporally limited world. It required revelation, Thomas thought, to know which is the case.[17]

In ancient Greek and medieval cosmology, there is a distinction between physics and ethics. But physical structure and moral meaning are aspects of one and the same world. The geocentric universe is suffused with the purposeful motion that comes from divinity. There is, however, a first fissure between structure and meaning. Aristotle's prime mover, the source of meaning, is not part of the world. It resides beyond the sphere of the fixed stars. And in medieval theology the distinction between God and the world is sharper still.

THE DIVISION OF MORAL COSMOLOGY INTO PHYSICS AND ETHICS

Yet the real wedge between physics and ethics came with the abandonment of the geocentric universe. What inspired Nicolaus Copernicus (1473–1543) was not just a new vision of the world, but even more so a revolutionary emphasis on explanation. The point of explanation, as Copernicus understood it, was not the fullness but the simplicity of a vision. What offended Copernicus were the incongruities in the cosmology of his predecessors.

> With them it is as though an artist were to gather the hands, feet, head and other members for his images from diverse models, each part excellently drawn, but not related to a single body, and since they in no way match each other, the result would be a monster rather than a man.[18]

But it was not really an organism like a human being that was to serve as the Copernican alternative to the geocentric monstrosity. It was a mathematically describable mechanism. What bothered Copernicus was the "uncertainty of mathematical tradition in establishing the motions of the system of the spheres" and the inability of philosophers to "agree on any one certain theory of the mechanism of the Universe, wrought for us by a supremely good and orderly creator."[19]

We can see here how physics and ethics are tending away from one another while nominally still of a piece. Cosmology is entrusted to "the philosophers" who have yet to split into physicists and ethicists. The "mechanism of the Universe" is still tied to goodness and order through the creator, but the norms of goodness are no longer inscribed in physical reality as they were for Thomas. Copernicus once more invoked the commanding presence of an animated universe when he pleaded for the central position of the sun.

In the middle of all sits Sun enthroned. In this most beautiful temple could we place this luminary in any better position from which he can illuminate the whole at once? He is rightly called the Lamp, the Mind, the Rule of the Universe; Hermes Trismegistus names him the Visible God, Sophocles' Electra calls him the All-seeing. So, the Sun sits as upon a royal throne ruling his children the planets which circle round him. The Earth has the Moon at her service. As Aristotle says, in his *de Animalibus*, the Moon has the closest relationship with the Earth. Meanwhile the Earth conceives by the Sun, and becomes pregnant with an annual rebirth.[20]

But this is retrospective rhetoric. It uses the images of the old cosmology to plead for the new. Copernicus immediately goes on to underscore the scientific and explanatory power that lends cogency to the heliocentric theory: "So we find underlying this ordination an admirable symmetry in the Universe, and a clear bond of harmony in the motion and magnitude of the orbits such as can be discovered in no other wise."[21]

Copernicus, Galileo, Kepler, and Newton are the names that are commonly attached to the stages of early modern cosmology. Galileo Galilei (1564–1642) adopted, elaborated, and advocated the Copernican system. It was the invention of the telescope in Holland that opened up the particulars of heaven. Galileo could see that the Moon was not made of ethereal stuff, that Venus had phases like the Moon, that Jupiter had satellites and the satellites had eclipses, and much more. Many of these observations made sense only on the assumption of a Copernican system.

Galileo's mature cosmology collided with the professoriate before it clashed with the Church. Philosophers at the time were still in charge of physics and beholden to Aristotle or Plato. The Church eventually sided with the philosophers and condemned Galileo's cosmology. The tangle of claims, charges, denunciations, and condemnations is an interesting confusion of the religious or moral and the scientific or physical. The philosophers and the Church objected to Copernicus and Galileo on both scriptural and physical grounds. Galileo and more enlightened Churchmen proposed as had Augustine, in so many ways a modern before his time, that the scriptural remarks on cosmology should be taken metaphorically. This has in fact become the standard way reasonable religious people read their scriptures.

The reasonable way is often called the liberal way, and liberalism in religion is scorned both by religious fundamentalists and by atheists. Both admire what they think to be forthright and substantive claims. Liberals are thought to be wishy-washy and spineless. Thus, a definite atheist like the physicist Steven Weinberg claims to prefer the fundamentalists to the liberals. "Religious Liberals," he has said, "are in one sense even farther in spirit from scientists than are fundamentalists and other religious conservatives. At least the conservatives like the scientists tell you that they believe in what they believe because it is true, rather than because it makes them good or happy."[22]

Weinberg's critique must disquiet liberals not just because here a reasonable man "in one sense" agrees with their religious opponents, but even more so because he assumes what liberals have a hard time disputing, that when ethics is severed from physics, truth goes with physics, and ethics is left with arbitrariness or at best with mere personal preference. On that assumption a moral cosmology can never be more than the best current physics (having the best claim on truth), embroidered with moral predilections.

A less conspicuous, but equally powerful contribution to the divergence of physics and ethics was Galileo's contribution to mechanics. The reassuringly ordered and animated Aristotelian cosmology had it that more massive things seek their natural place more strongly, i.e., they fall more rapidly, and that lateral motion requires a sustaining force as anyone knows who has pushed a wheelbarrow and was glad to set it down. Galileo recognized that without intervention motion continues indefinitely. Though he probably did not drop rocks of different masses from the leaning Tower of Pisa, Galileo did have balls roll down inclined planes—a case of slowed-down drops. He discovered that the distance traversed was proportional to the square of the elapsed time regardless of the mass of the ball and the angle of the plane.[23]

Galileo made his case geometrically and in the mathematical language of his time. The language of algebra that allows us to state laws of nature more generally and abstractly was just being developed during Galileo's lifetime. But the underlying question of whether mathematics can capture the variety and complexity of reality was explicit to Galileo.[24] However, mathematics was for him not just a means of description. Somehow simply getting the right results through mathematics was neither here nor there.[25] Nature, rather, had an underlying mathematical structure, and without the rigor of mathematical language, descriptions and explanations of reality had to remain superficial and aimless.

Although it is simply wrong to claim that light things fall more slowly than heavy ones, a feather does descend in a leisurely way while a rock drops decisively. But the ways things and persons disclose their character in their movements are compounds of many and distinct properties. We think of a feather as falling softly because it is itself soft, has once been part of a fragile bird, has a gentle color, and of course is much more exposed to the friction of air than a rock. All these enchanting contingencies and particulars drop away when the true nature of motion is at issue.

Similarly, Galileo searched for the lawful structure of matter that in everyday reality addresses us as hard, rough, cold, and black or as soft, smooth, warm, and gray and in a numberless variety of other textures. In Aristotle's world, matter basically comes as the four elements. Galileo, however, searched for one basic theory of matter that would explain the appearance of things (their secondary qualities, so called) as the superficial impressions that are left on the human senses by the basic structure of matter. The moral eloquence of things is fading. But there is a reward for seeing the true nature of things—peace of mind.[26]

Such restful insight recalls Aristotelian contemplation and propelled Johannes Kepler in his cosmological investigations. Christian theology seems to have been a major impetus. Kepler wanted to understand how the cosmos discloses the trinitarian God. Although this challenge was ever before his mind, Kepler's Trinity is the God of the philosophers, not the God of Abraham, Isaac, and Jacob, as Pascal would have put it.[27] God was the origin of the cosmic structure, and to that God, as Heidegger had it, "humans can bear neither prayers nor sacrifices. Before the *Causa sui* [the ultimate uncaused cause] humans can neither humbly genuflect nor can they, before that God, make music and dance."[28]

Kepler's cosmology is aesthetically striking and would be logically compelling if only it were correct. Kepler wanted to know why there were exactly six planets, why not more or fewer. His answer began with the five regular solids, the ones whose surfaces are composed of equilateral and equiangular faces, triangles, or squares, or pentagons. There are five and only five such—the perfect or Platonic solids. If you imagine spheres enclosing and being enclosed by these polyhedra, then, if you nest them the right way, you discover why there are just six planets and why their orbits are just at their particular distances from one another. Here is Kepler's succinct account:

> *The Earth is the circle and measure of all: To it circumscribe a Dodecahedron: The Circle comprehending that will be Mars. To Mars circumscribe a Tetrahedron: The Circle comprehending that will be Jupiter. To Jupiter circumscribe a cube: The Circle comprehending that will be Saturn. Now then to Earth inscribe an Icosahedron: The circle inscribed to it will be Venus. To Venus inscribe an Octahedron: The Circle inscribed to it will be Mercury.*
> You have the reason for the number of planets.[29]

Kepler himself came to recognize that this picture is only approximate and had to be supplemented if not abandoned. His legacy, in any case, is not the formally enchanting cosmos, but three laws of motion that are of mathematical character. The first states that the planetary orbits are elliptical, not circular. The second says that a line from a planet to the sun sweeps equal areas in equal amounts of time. Thus, the second law captures the increase in orbital velocity as a planet comes closer to the focus of the ellipse where the sun is located. As the line to the sun gets shorter, the planet has to move faster so that the shorter line sweeps the same area in the same time interval as it does when the line is long. The third law tells us, roughly put, how the time it takes a planet to circle the sun (its period) is related to its distance (its semi-major axis) from the sun: The square of the period is proportional to the cube of the distance. Thus, to put it roughly once more, as you move outward from the sun, the periods of the planets increase greatly, but the distances from the Sun increase even more. The celestial phenomena that Kepler's laws refer to were more or less familiar long before Kepler. His genius lay in his ability to grasp them mathematically, and it was his contribution to mathematical physics rather than to aesthetic cosmology that we remember him for.[30]

Isaac Newton's *Mathematical Principles of Natural Philosophy* (1687) is the triumphant conclusion of early modern cosmology.[31] Newton unified Galileo's laws of terrestrial and Kepler's laws of celestial motion into his three laws of motion and the law of universal gravitation. Like Euclid, Newton began his *Principles*, commonly referred to by its original Latin title as *Principia*, with definitions and followed them with "Axioms, or the Laws of Motion."

Law 1: Every body perseveres in its state being at rest or moving uniformly straight forward, except insofar it is compelled to change its state by forces impressed.

Law 2: A change in motion is proportional to the motive force impressed and takes place along the straight line in which that force is impressed.

Law 3: To any action there is always an opposite and equal reaction; in other words, the actions of two bodies upon each other are always equal and always opposite in direction.[32]

It's hard for us to imagine what cultural impact Newton's theory had in his time. The educated classes of Europe thought it their duty to become familiar with the *Principia*. The theory was widely discussed, admired, and applied. Portraits and statues of the great man appeared everywhere. Newtonian physics made for a perspicuous and parsimonious explanation of cosmic matter and motion. Mechanics is often defined that way: the science of matter and motion. Due to Newton's accomplishments, mechanics became the queen of the sciences and very nearly of knowledge simply. As Ernest Nagel put it, "by the middle of the nineteenth century mechanics was widely acknowledged as the most perfect physical science, embodying the ideal toward which all other branches of science ought to aspire."[33] It seemed to be "the only alternative to obscurantist philosophy."[34]

In the Preface to the first edition of the *Principia* Newton himself championed rational or universal mechanics as the master science whose rule might cover all natural phenomena.

> If only we could derive the other phenomena of nature from mechanical principles by the same kind of reasoning! For many things lead me to have a suspicion that all phenomena may depend on certain forces by which the particles of bodies, by causes not yet known, either are impelled toward one another and cohere in regular figures, or are repelled from one another and recede.[35]

If we think of the universe as a machine, Newton's mechanics tells us how the machine works, but, as Newton well saw, it does not tell us how it came to be. Newton's "mechanical principles" are formulated as his laws of motion and gravitation. But lawfulness alone explains nothing. Why did an apple fall on Newton's head (if in fact it did)? The law of gravity tells what force existed between the apple and the earth, but for the apple to come down on Newton's head certain conditions had to be met. The center of the earth, Newton's head, and the apple must have been so arranged that a straight line went through all three of them, and the apple must have come loose from its bough. These contingent conditions need to be added as premises to Newton's laws for an explanation of an actual event to follow.

Contingency, then, is the complementary opposite of lawfulness. Laws are universal and necessary; contingency is particular and factual. But does inquiry simply come to a halt when it runs into the brute facts of contingency? Can't we somehow explain contingency? There are two ways of accounting for contingency by simply explaining it away. Perhaps the arrangement of particulars is the way it is because it could not possibly be otherwise. It's the way it is because it's the only way it can be. Contingency is just necessity in disguise. Kepler had searched for something like that solution when he proposed that the arrangement of the planets was governed by the structure of the regular solids. There are no more and no fewer than five, and each assigns an orbit to a planet in relation to its neighboring planet and to the sun. The second way of dissolving contingency is to derive it from randomness. To ask for an explanation or derivation of randomness itself is to misunderstand randomness. If an arrangement shows a rhyme or bespeaks a reason, it's not random. If contingency is the offspring of randomness, it's unnecessary and in fact impossible to explain it.

A third way of answering curiosity about contingency is to show that curiosity misses the force of contingency, or at least the kind of contingency that animates the universe—the presence of divinity. The fundamental attitude of the Blackfeet and the Peoples of the Book toward the universe was piety, and for Aristotle it was contemplation. There was curiosity, of course, about this or that aspect of the world, but the world as a whole was met with reverence.

Kepler saw divine presence in the soul that moved the stars and the comets (and that he later replaced with "force"). Newton made a similar suggestion in his efforts to understand the force of gravity. The law of gravitation that posited attractive forces emanating from matter and acting instantaneously over distances however great looked to contemporary critics, to Leibniz in particular, like a violation of mechanics and lapse into obscurantism. Newton admitted that he had "not yet assigned a cause to gravity" and that he had "not as yet been able to deduce from phenomena the reasons for these properties of gravity." Still, he was satisfied "that gravity really exists and acts according to the laws that we have set forth and is sufficient to explain all the motions of the heavenly bodies and of our sea."[36]

Yet the question of the causal nature of gravity continued to occupy Newton, and in the Appendix to *Opticks* (4th ed. 1730) he ventured a religious or at least metaphysical explanation. He rejected the idea that gravity might be transmitted by a medium such as a "dense Fluid."[37] He suggested instead that, just as the motions of an animal follow from its center of awareness, its "sensorium," so infinite space might be the sensorium of "a Being incorporeal, living, intelligent, omnipresent."[38] And the order and motions of the heavenly bodies follow from the divine sensorium? That obviously is the suggestion though Newton did not say so explicitly.

Newton's explicit view of contingency envisioned a definite but distant divinity. God orders the contingent structure of the universe not through divine presence but through a divine design. Again, it's the arrangement of the planets that challenges our understanding. Newton wondered how it came about that "[the] six primary planets revolve about the sun in circles concentric with the sun, with the same direction of motion, and very nearly in the same plane." His answer in the *Principia* is: "This most elegant system of the sun, planets, and comets could not have arisen without the design and dominion of an intelligent and powerful being."[39]

Newton did not think that the cosmic order could have arisen "from blind metaphysical necessity," and perhaps we can read his remarks as the rejection both of the blindness of randomness and of the force of logical or mathematical necessity. Necessity is a property of God. It's important to note that God, for Newton, was more than the designer of a machine that, once designed, is left to its own devices. "For we worship him," Newton said at the very end of the *Principia*, "as servants, and a god without dominion, providence, and final causes is nothing other than fate and nature."[40] For Newton this point was more than a personal aside. He concluded his discussion of God with the claim that "to treat of God from phenomena is certainly a part of 'natural' philosophy."[41] Though Newton has us worship God as servants, the moral force of divinity has become thin. God has become an abstract physical or metaphysical issue as Pascal and Heidegger have pointed out.

In the history of ideas, it was Immanuel Kant (1724–1804) who clearly saw the end of moral cosmology. As a young scholar he both applauded and undermined Newton's thin moral cosmology. In his *Universal Natural History and Theory of the Heavens* of 1755, he saw in "the wisdom, the goodness, the power that have revealed themselves [in the universe]" the hand of a divine designer.[42] At the same time he conjectured that we can derive the structure of the universe from Newtonian lawfulness if we posit an initial state of random or chaotic distribution of atoms in infinite space. In his mature philosophy, most notably in the *Critique of Pure Reason* (1781), he rigorously rejected the idea that the heavens declare the glory of God. For Kant, space was not, as Newton suggested, the sensorium of God, but a form of the human sensorium. Divinity disappeared from the world, but ethics, of course could not. It retreated from objective reality to the human subject as is evident from Kant's memorable saying (as discussed in chapter 1) that concludes the *Critique of Practical Reason* (1788): "Two things fill the mind with ever new and increasing admiration and reverence, the more often and steadily our reflection attends to them, the starry heavens above me and the moral law within me."[43] Moral cosmology had fallen apart into physics and ethics. Almost two centuries later, in 1976, Martin Gardner put the matter more prosaically and explicitly:

> If the reader wonders why the book *[Relativity Simply Explained]* contains no chapter on the philosophical consequences of relativity, it is because I am firmly persuaded that in the ordinary sense of "philosophical," relativity has no consequences. For the theory of knowledge and the philosophy of science, it obviously has implications, chiefly through its demonstration that the mathematical structure of space and time cannot be determined without observation and experiment. But as far as the great traditional topics of philosophy are concerned—God, immortality, free will, good and evil, and so on—relativity has absolutely nothing to say.[44]

There is a story that poignantly pictures the fall of moral cosmology. On May 11, 1870, Charles R. Sunderlee wrote a letter from Old Militia Camp on the Yellowstone to the *Helena Daily Herald* in Helena, Montana Territory.[45] His party encountered a band of Shoshone-Bannocks "who would not live with their brethren, in peace with the whites; but who prefer living remote from all Indians, and civilized beings; foes of their former tribes and of the whites." They stole three packhorses, and when pursued, tried to cross the Yellowstone River on "a hastily constructed raft" just above the Yellowstone Falls. In spite of their best efforts, they kept drifting toward the Falls. "When about fifty yards below where we were standing," Sunderlee wrote, "an old Indian arose and stood erect in the center of a circle of Braves, he spoke a few words, turned his face towards the sun and seemingly bade it farewell, then wrapping his robe around him, he sat down." The others, "seated in a circle shook hands and then commenced wailing their always mournful death song."[46]

NOSTALGIA FOR A MORAL COSMOLOGY

We lack a moral cosmology and seem to be unable to master the inevitable first step toward a comprehending and comprehensive cosmology, that is, to grasp the incisive and rigorous cosmology that modern physics has uncovered. Yet we are reluctant to give up on what used to anchor moral cosmology. We are suffering from an incurable nostalgia for a centered, animated, and storied world. To call something nostalgic these days is to dismiss it as lacking in vigor and relevance. That's how social theorists use the term. To ordinary folks, nostalgia feels like a subliminal ailment.

The term "nostalgia" was coined in 1688 by the Swiss doctor-to-be Johannes Hofer.[47] He observed a pathological syndrome of depression and listlessness among young people who had left or had been taken from their native home. He described the syndrome as consisting of the following features:

> The diagnostic signs that indicate imminent Nostalgia are to be searched for in the very condition of the young men: for if they often walk about sad, if they turn aside foreign customs, if they are disdainful of conversations about foreign matters, if their disposition inclines them to melancholy, if they react painfully to jokes or the slightest injuries or to other unpleasant issues, if they eagerly and often lapse into talking about the pleasures of the fatherland and prefer them to foreign ones, and other matters of this kind: one certainly has to judge these men to be extremely disposed toward Nostalgia.[48]

Hofer noted that the Swiss seemed particularly prone to this disease and more particularly those from the Berne area. Napoleon's Swiss mercenaries were supposedly driven to melancholy, desertion, or suicide when hearing a tune, rendered as Kühe-Reyen (cow dance) in Hofer's dissertation. Heidi, of course, is the embodiment of Swiss homesickness for many of us.

Nostalgia seems rare these days though it gravely afflicted the Navajo who, following the Navajo-Hopi Land Settlement Act of 1974, were forced off their ancestral lands.[49] Even before the violent disruptions of colonization, homesickness was not unknown in Native American culture. In the founding story of the Blackfoot Sundance, So-at-sa-ki (Feather Woman) who had married Morning Star and lived in the sky, "dug up the sacred turnip, creating a hole in the sky," as James Welch's novel *Fools Crow* tells us. "She looked down and saw her people, her mother and father, her sister, on the plains and she grew homesick."[50] Her nostalgia is the inversion of the kind we suffer today. Most of us have shallow roots in our hometown if in fact there is a place we can call that, and so we suffer little when we move or are moved. Hence the change of meaning that "nostalgia" has undergone. As Edward Casey has pointed out, the longing of nostalgia has been diffused and displaced toward the centered, animated, and storied universe we no longer inhabit.[51]

In the Western world, the culture of Christmas is a reminder of how on occasion we feel drawn to the picture of the well-ordered world. The Christmas cards we pick to convey our sentiments often portray a pretechnological village setting—snowy fields, shingled barns, horse-drawn sleighs, candlelit homes, churches illuminated from within, carolers going house to house, the manger scene suffused with the promise of hope there and then. Much of this is sentimental, of course, most of it is commercialized, and some of it is insensitive to people who do not believe in Christ or do not believe in God.

Törbel is a village that answers our nostalgic longings. It's located in the Swiss Alps on a high bench, roughly northwest of the Visp valley that famously starts in Zermatt at the foot of the Matterhorn.[52] The village of about six-hundred souls sits at 4,500 feet, a cluster of houses and barns built of larch timbers, blackened by the sun. There are vegetable gardens around the houses, wheat fields and irrigated hay meadows below them, pastures and forests above them, and, crowning it all, the rock and ice of the Alps. The entire area is veined with irrigation ditches and crisscrossed with dirt roads and paths. These several areas and lines are the outcome of about a thousand years of clearing, digging, mowing, plowing, and grazing. The area has the character of a face that is lined with the traces of labor, sorrow, and pleasure.

Yet there is more than character. There are focal points of space and time that provide for orientation. At the center of the Törbel region is the church, surrounded by dwellings and barns. Christmas and Easter are the great feast days of the liturgical year, Christmas falling in a time of quiet and leisure, when people live off the wheat and cheeses of summer and the cattle are sheltered in barns and winter on hay. Easter comes in spring, a time of repairing the irrigation ditches, clearing the pastures of debris, and plowing the wheat fields.

Such is Törbel, or rather that's the way it was a century ago and essentially had been for half a millennium before. Industrialization began to invade Törbel after World War I and transformed it after World War II. In its preindustrial shape, Törbel was unusually close socially and self-sufficient economically even among European villages. Like all of them, however, it was more and more exposed to the surrounding power and prosperity of technology. A motorized mowing machine allowed a person to cut four times as much grass as it took a person with a scythe. Working in a factory down in the valley allowed one to buy things that a farming relative could only dream of.

Today alpine agriculture is a historic artifact, anxiously kept alive by the government. Economically, Törbel is a bedroom community and a base for the tourist industry. The Christmas card past is captured by the writings and pictures of Helen Güdel.[53] She lives in Törbel, but it's telling that Törbel is not the place where she was born and raised. It attracted her the way we may be captured by her work—wintry scenes of the traditional houses huddled together, populated by cows, goats, and horses, with some evidence of machinery and technology such as power poles and wires.

Our nostalgia is incurable because Dr. Hofer's remedy, "a return to the homeland" that is also a return to a moral cosmology, is no longer available. There is still a geographical spot called Törbel, and the charms of its past are still palpable. Yet life in Törbel is no longer centered on the church, animated by the struggle with nature, and storied by the sermons on Sunday and the tales handed down from elders to children. Yet if nostalgia is not remediable, neither is it eradicable. Our evolutionary roots and cultural legacies keep us tethered to a position that once extended into a moral cosmology, but now seems to cripple us when we're challenged to appropriate conceptually, if not also mathematically, contemporary astrophysics. If we can't cure our nostalgia and get at least a rudimentary conceptual and mathematical grip on the cosmology of astrophysics, there is no prospect of a moral cosmology. There is of course no sense in trying to find one if physics and ethics are disjoint and knowledge of the cosmos has no moral point.

If a moral cosmology is to be possible, laypeople and moral philosophers particularly need to get, as one of the first steps, a conceptual and mathematical grip on astrophysics. Physicists need to make cosmology clearer to themselves and clear to the person in the street or in the bar. In an apocryphal remark, rendered in many versions, Nobel physicist Ernest Rutherford is supposed to have said that if you can't explain your theory to the bartender, you don't understand it yourself. And the celebrated Stephen Hawking held out the hope that "if we do discover a complete theory, it should in time be understandable in broad principle by everyone, not just a few scientists."[54]

We would then have at least a basis for a moral cosmology. But what is or what was a moral cosmology like and how did we lose or abandon it? We use "biology" in a generic and in a specific sense. Generically, biology is the broad and varied science of life. Specifically, it is a domain of life or the particular study of a certain phenomenon, as in "The Biology of the Honey Bee" or "A Biology of Intentionality." Similarly, I will use "moral cosmology" as a generic term for discussions of the question of whether cosmology has a moral point, and I will use "a moral cosmology" to refer to a particular conception or theory regarding the moral significance of cosmology.

One other factor needs to be considered to situate the romantic response to the mechanical universe. The romantics were not the advocates of a universally realized need. Not everyone felt a grievous loss in the face of the rising modern era. There was always one party or another that was confident of rational explanation without remainder and of beneficial reconstruction from the ground up. All that was needed was supposedly available—building materials (atoms) and rational designs (scientific and technological theories). The constituencies and banners of this movement have shifted and changed over time, but it is if anything more vigorous and self-confident now than it was some two hundred years ago.

THE ROMANTIC SEARCH FOR A MORAL COSMOLOGY

"Two things fill the mind with ever new and increasing admiration and reverence, the more often and steadily our reflection attends to them," Kant said famously in the conclusion to the *Critique of Practical Reason* (1788), "the starry heavens above me and the moral law within me."[55] Ever since, moral cosmology has been unavailable as a backdrop for daily life and for scholarship in the humanities. Kant's dictum did not, of course, signal the end of cosmology simply, not even for Kant. The cosmos, after all, continued to fill his mind "with ever new and increasing admiration and reverence." Kant and philosophers in general, moreover, are careful observers rather than revolutionary agents of their culture.

Beginning with Copernicus, cosmology had been developing into a chiefly scientific discipline and has remained such to this day. What has been lost is the moral force of cosmology that would inform our sense of what it is to be human and our awareness of the place we occupy in the universe. As a consequence, perhaps, philosophers have for the most part lost interest in cosmology or, more accurately, have surrendered it to the sciences. Cosmology failed to rate an entry in the 1995 Cambridge *Dictionary of Philosophy*.[56] In the 1998 *Routledge Encyclopedia of Philosophy*, Ernan McMullin distinguishes between cosmology as a worldview, cosmology as a brand of metaphysics, and cosmology as a scientific enterprise, devoted to "the construction of plausible universe models."[57] The first two kinds of cosmology are mentioned only to be dismissed. On September 26, 2017, Christopher Smeenk and George Ellis published an entry in the *Stanford Encyclopedia of Philosophy*, titled "Philosophy of Cosmology."[58] Here again the emphasis is entirely on scientific cosmology even though the entry concludes with an insightful paragraph on what the authors call *cosmologia*, a close neighbor of moral cosmology.

When I speak of the search for a moral cosmology, I take "moral" in the wide sense of Hume and Dewey to include not only modern ethical norms like rights, pleasure, and tolerance, but also supererogatory norms like heroism, generosity, and diligence, and what are now considered nonmoral norms such as virtuosity, vigor, or grace. "Moral" in this broad sense refers to all the excellences that characterize the culture of a particular time and place. Thus, a moral cosmology would give us an encompassing sense of orientation and of the good life.

Mythic cosmology was surely moral in this sense. For the ancient Greeks as much as for the Blackfeet in what is now Montana the world was a spiritual plenum, resonating with the voices of divinities, both with those of plants, animals, streams, and mountains and with the voices of the Sun, the Moon, the planets, and the stars. It was a world to which humans responded in a spirit of piety. To be pious was to be attentive to the commands of gods and spirits and, in the end, to be favored with grace and good fortune. Virgil's Aeneas, *pious Aeneas*, is an exemplar of such piety.[59]

Thales may have been the first to be moved by curiosity rather than piety.[60] Not that everyone was pious prior to Thales. Blasphemers were not, but blasphemy is the tribute impiety pays to divinity. Nor were people before Thales lacking in curiosity.

Hunters and peasants had to be curious to survive, and geometers had, by Thales's time, shown disciplined and ingenious curiosity. But Thales was the first to be moved by cosmic curiosity, he was the first to think that laws rather than gods ruled the universe. To be sure, Thales's laws of condensation and rarefaction were rudimentary and closely tied to the stuff they governed.

In Aristotle's physics, there are more distinct laws of motion.[61] But they too are linked to the divine order of the universe. We can see here a magnificent, if unsustainable as it turned out, balance between the moral resonance and the physical lawfulness of the cosmos. What was almost seamlessly one in mythic cosmology, Aristotle unified into an artful structure of three elements—metaphysics, cosmology, and ethics.[62] I will use the notion of scientific lawfulness as a thread to follow roughly the long and complicated developments that led from Aristotle's moral cosmology to the search of the romantics and beyond.

To begin, then, with Aristotle and to speak informally, things in Aristotelian ontology had a certain character. They were fiery, heavy, or ethereal, and they moved in characteristic ways—fiery things tended upward, heavy things tended downward, and ethereal things, the planets and stars, moved in perfect circles with perfect uniformity. In Galileo's kinematics, almost two millennia after Aristotle, things begin to lose their character. Light things tend to fall no differently from heavy things. Cognitive interest shifted from the character of things to the uniform laws that govern their behavior. This shift reached a first point of resolution and rest in Newton's laws of motion that resolved Galileo's laws of terrestrial with Kepler's laws of heavenly motion into one encompassing theory.[63] Things that once had a color and personality of their own were now reduced to point masses and mere instances of the laws that covered them.

This ontological development converged with changes in astronomy that began with Copernicus. The naive geocentric vision of the world has an engaging solidity and intuitive plausibility. But it could only be sustained, in view of careful astronomical observations, at the price of the extravagantly complex Ptolemaic astronomy. Copernicus exchanged terrestrial solidity and astronomic complexity for cosmic simplicity and intuitive implausibility. Here again Newton's *Principia* is the endpoint of a development, the rigorous explication of the structure of astronomy through physical lawfulness.[64]

Once things had lost their commanding presence, the moral eloquence of reality had to fall silent as well. A natural-law-physics leaves no room for a natural-law-ethics. Morality was lost to reality and became a matter of pragmatic prudence in Descartes.[65] It entirely detached itself from reality in Hume's is-ought distinction.[66] It was finally set over against knowledge and reality in Kant's ethics of practical reason, an opposition reflected in the Kantian dictum quoted at the start of this section.

Yet it remains that no one has been able to answer the romantic complaint that there is more to the world than a mechanical universe and a mercenary world and that we cannot plumb the depth of our present predicament, far less look forward to a moral cosmology, and that we cannot be fully human beings, until the missing regions of reality have been recovered by an appropriate philosophy and appropriated by vigorous practices. To be sure, the mechanical universe has yielded to astrophysics and the Philistine society to the affluent society; but the fundamental cosmological predicament is much the same today as it was at the turn from the eighteenth to the nineteenth century.

To understand the romantic concerns, it is helpful to see them against the background of a rival philosophy that took the Industrial Revolution seriously, unlike the haughty neglect of the romantics. Marx and Engels were critics of the dawning industrial culture. As they observe in *The Communist Manifesto*:

> Subjugation of the forces of nature, machinery, application of chemistry to industry and agriculture; steam-powered navigation, railroads, electrical telegraphs, settlement of entire world regions, making rivers navigable, entire populations generated from the bottom up—which prior century could have foreseen that such productive forces could have slumbered in the bosom of social endeavors.[67]

Marx and Engels not only saw the rise of industrial power, they also foresaw the overwhelming production of goods, "the epidemy of overproduction." But in their single-minded and erroneous conviction that a revolution was inevitable, they saw in overproduction a mortal crisis of the bourgeoisie. "The bourgeois circumstances," they said, "have become too narrow to capture the prosperity they have produced." They continued:

> How does the bourgeoisie overcome these crises? On the one hand through the inevitable destruction of a mass of productive forces; on the other hand, through the conquest of new markets and the thorough exploitation of old markets. How in short? Through the preparation of more encompassing and powerful crises and the decrease of the means of preventing the crises. The weapons, whereby the bourgeoisie beat feudalism into the ground, are now directed at the bourgeoisie itself.[68]

Romantics, as I understand them, hold that you cannot truly look forward without looking back. What needs regard in the romantic view are the great continuities of culture and nature. It is against the cultural and natural background that the real liabilities and possibilities of the present come into view. Those theses need substance and detail, of course, and I will now try to furnish these by turning to Kant, Goethe, and Schelling and using the lawfulness of nature as the pivot of my investigation.

RATIONAL COSMOLOGY—KANT

Cosmology was an early and prominent concern of Kant's. His first ambitious work, published anonymously in 1755 and immediately the victim of his publisher's bankruptcy, is the *Universal History and Theory of the Heavens or Essay on the Constitution and Mechanical Origin of the Entire Cosmic Edifice, Carried Out According to Newtonian Principles*. As the title shows, Kant's treatise is remarkable for its wholehearted embrace of Newton's mechanical lawfulness. In historical context, however, it is equally notable for the moral inferences Kant drew from his cosmology.

Though Newton famously explicated the structure of the solar system through his laws of motion, he was unwilling to attempt an explanation of how that structure had come into being. Newton thought that "this most elegant system of the sun, planets, and comets could not have arisen without the design and dominion of an intelligent and powerful being."[69] Kant employed Newton's mechanics beyond and against Newton to propose an explanation of how the universe evolved from an original chaos of atoms in infinite space. Though many of his explanations turned out to be wrong, Kant delineated a crucial trend in the devolution of moral cosmology.

To illustrate the issue, consider the question why the sun and its planets constitute a stable system. Kant attempted an explanation in terms of attractive and repulsive forces that led by coagulation and according to differences in density to planets orbiting at appropriate distances to the sun and in the same plane with the sun. This explanation is wrong in many details though it does agree in spirit with current theory.

Regardless of these errors, the moral point at issue is of general significance. In Newton's astronomy, there is a definite local indication of divine action, an internal element, or more accurately, a fragment of moral cosmology since, even if Newton were correct, the sharply defined and localized divine presence within cosmology would do next to nothing in teaching us what place we occupy in the world and how that placement tells us who we are and what we are to do. At any rate, the subsequent scientific explanation of planetary stability dissolves God's presence within cosmology.

There is a quaint piece of internal or local ethics in Kant's cosmology. It derives from his thesis that warm climates make people think and act slowly and cold ones quickly and wisely. Hence the intelligent creatures on Jupiter likely possess more wisdom than the slow and dim-witted inhabitants of Venus. Humans are so placed in the solar system that they occupy the perilous middle position "where the temptations of sensuous stimulations possess a powerful capacity of seduction against the supremacy of the spirit and where the spirit nevertheless cannot deny the ability whereby it can resist those temptations."[70]

The prevailing moral import of Kant's early cosmology is, however, universal or systematic. The moral significance of the cosmos is not to be found in a particular location or feature of the universe but rather in its entire order or structure. It is, moreover, of a theological cast. Commenting on the incomprehensible infinity of the cosmos, Kant said: "The wisdom, the goodness, the power that have revealed themselves are infinite and to the same extent fruitful and active; the design of their revelation must therefore, likewise be infinite and without limits."[71] Kant attempted to build a bridge between his internal and his systematic ethics by arguing that humans must aspire to the spiritual dimensions humans owe to their distance from the origin and center of the universe if they are to grasp the systematic import of cosmic morality.

Kant's systematic moral cosmology evidently has more practical force than Newton's local moral cosmology. If Kant is right, we know ourselves to be placed in a world that reflects and witnesses to a wise, beneficent, and powerful God. Newton's cosmic morality, though practically bland, is more definitely and indisputably (if

Newton had been right) evident in astronomy. Thus, the best is divided between two worlds—Newton's indisputable and Kant's inspiring divine presence; and so being divided each is fatally weakened—Newton's by vacuity and Kant's by its tenuous attachment to astronomy, as the later Kant was the most vigorous to argue.

In the *Critique of Pure Reason* (1781) Kant denied that the universe exhibits any empirically demonstrable evidence of divine origin or design. Any attempt to pronounce conclusively on the ultimate constitution of reality, i.e., on physics, or on the all-encompassing structure of the universe, astronomy, leads to antinomies—to every apparently sound claim there is an equally persuasive counterclaim. Such fundamental or universal assertions, Kant argued, assume what we cannot have, knowledge of reality in itself and by itself. In particular, proofs for the existence of God are based on inconclusive evidence at best and must in any case illicitly slide (via the "ontological proof") from something that can be thought to something that supposedly exists in fact.

All this suggests that in the aftermath of the first *Critique* not only moral cosmologies are ruled out as far as Kant is concerned, but any cosmology that presents a conclusive physics and astronomy. What is left is cosmological research without cosmological conclusions. In the discussion of the antinomies, Kant's sympathies lie with the skeptical and empirical side. He applauds its insistence on an ever-open horizon of inquiry when it comes to the composition and origin of the world (though he rejects the skeptical denial of undetermined freedom and of the hypothetical or regulative use of metaphysical notions).

More to the point, the *Critique of Pure Reason* severed Kantian ontology both from cosmology and morality, and, as it turned out, this jejune theory of reality left Kant himself unsure and dissatisfied. The burr under the saddle turns up when we look more closely at the connection between Kantian and Newtonian ontology. In the conventional view, Kant attempted to justify Newtonian physics by deriving it from synthetic a priori principles. I agree, however, with Gordon Brittan that Kant argued for more and less than the common interpretation has it—less in arguing, though not without inconsistencies, that Newton's physics is merely one among indefinitely many "really possible" kinds of physics; more in that he gives Newtonian physics a realist rather than an instrumentalist or phenomenalist reading. It is Kant's realism that makes him search for a substantial ontology. Kant thought that Newton's laws of motion gave the best available account that actually conformed to the strictures of real possibility. He disagreed, however, with Newton's cosmological extension of mechanical lawfulness, viz., with Newton's claim that the universe is enclosed in absolute space and composed of impenetrable atoms in empty space.

Kant had originally conceived of the first *Critique* as an enterprise of clearing the ground whereon to erect a metaphysics that avoided the airy arrogance of the rationalists and the studied ignorance of the skeptics and instead had application to experience. He proceeded to do just this for the natural sciences in the *Metaphysical Foundations of Natural Science* of 1786.[72] What was conceived as an enterprise of secure and orderly construction came to reveal a problem Kant was finally unable to solve. It lay in the implicit question of whether the account of a mechanical universe adequately reflected the force of reality. Divine power had been reduced to a postulate of nonempirical ethics, and the ordinary force of things remained philosophically unaccounted for since it appeared only as the unpredictable content of the a priori conceptual forms or categories.

The romantics responded primarily to physics, and their response can be read as a critique of atomist and mechanist science. But we must not forget that "mechanical" is also, by etymology and semantics, the adjective that belongs to the noun "machine." It is one of the great ironies in the history of ideas that in the treatises, essays, novels, and dramas of what was between 1770 and 1830 perhaps the greatest flowering of German letters, the seminal development that in time entirely transformed terrestrial reality and by now has eclipsed if not destroyed the world of letters went almost unnoticed—the rise of machinery and technology in the Industrial Revolution.

THE FORCE AND THE LIMITS OF CONTINGENCY—GOETHE

Almost, but not entirely. Goethe in *Wilhelm Meisters Wanderjahre* has Susanne, who presides over a preindustrial spinning and weaving business, say:

> The overpowering rise of machinery pains and frightens me; it is rolling along like a thunderstorm, slowly, slowly; but it has taken its direction, it will come and strike.[73]

The romantics were not entirely unaware of the industrial and commercial transformation of reality.[74] They felt estranged from this powerful development, and their estrangement, more as a background condition than a focal concern, lent additional urgency to the search for a moral cosmology.

Before Susanne revealed her anxiety about machinery, her interlocutor noticed "a certain melancholy, an expression of concern" about her, and upon concluding her diagnosis, Susanne can only see a choice between two unlovely alternatives—joining the rush to machinery or emigration to America.[75] A similar despair clings sometimes to the notion of romanticism. When in social criticism and theory a position is called romantic, the appellation is often thought to be tantamount to a refutation— romanticism is hopeless and regressive if not reactionary. For German romanticism, this burden has been made worse by the truly reactionary, and worse than that, use the Nazis have made of romantic themes.[76]

While Kant is commonly thought to have tied his philosophy too closely to Newtonian physics, thus supposedly sentencing it to death, Goethe, in the conventional view, responded to Newtonian science with uncomprehending hostility. His *Farbenlehre* (doctrine of colors) is taken to illustrate his position best. Goethe could not get himself to accept Newton's demonstration that white light is the compound of the colors of the spectrum. Colors, Goethe claimed, were things that white light did and suffered. So strong was his opposition that Goethe pronounced his "refutation" of

"the error of the Newtonian doctrine" to be his epochal contribution to humanity.[77]
The same attitude appears where Goethe talked directly about astronomy. Cosmology
has an important place in Goethe's most important work, *Faust*. In the "Prologue in
Heaven," the archangels behold and praise the universe, but it is, alas, a geocentric
world, albeit with the inconsistency of the earth revolving rapidly about its axis.[78]
Immediate and thoughtful observation was Goethe's forte; the rigorous mathematical
approach to nature was foreign to him as he explained to Eckermann:

> In the natural sciences I have tried my hand in pretty much all areas; however, my con-
> cerns were always and only directed toward those subjects that surrounded me here on
> earth and that could be immediately perceived by the senses; which is the reason why I
> have never been concerned with astronomy since here the senses are not sufficient, but
> rather one needs to resort to instruments, computations, and mechanics which require
> a life of their own and were of no concern to me.[79]

When it comes to the structure of the universe at large, then, Goethe agreed with
Kant that the rise of Newtonian mechanics had emptied the cosmos of moral force
and interest. Goethe knew the solar system and its planets well enough, but it was their
immediate splendor rather than their lawful structure that engaged his admiration.[80]

Goethe was, however, deeply fascinated by a phenomenon of modern physics that
at the time was widely investigated and discussed under the headings of chemistry,
electricity, and magnetism. This was a period of excitement and frustration, excitement
about the rapid succession of discoveries, frustration about the scientists' inability to
explain and connect the discoveries in one lawful theory. Kant, in fact, thought that
chemistry would forever remain a collection of empirical observations that defied
mathematical methods and did not deserve to be called an "actual science."[81]

The phenomenon of elective affinity or attraction, so called, was as fascinating
to the imagination as it was recalcitrant to explanation.[82] To put it schematically,
a certain chemical compound of two elements, say AB, when mixed with a third
element, say C, reacts in such a way that A lets go of B and then combines with C
to yield AC. Thus, A and C have an elective affinity with each other that neither A
nor C has with B.

A still more fascinating case is crosswise elective affinity. Goethe has the captain
in the eponymous novel *Elective Affinities* give an example:

> Think of an A that is intimately bound to a B and cannot be separated from it by many
> different means and not even through some violence; think of a C that is likewise related
> to a D; now bring those two pairs into contact: A will throw itself on D and C on B,
> and you could not tell which first deserted the one and which first connected itself again
> with the other.[83]

To call this "elective affinity" is to explain nothing as Schelling was to stress repeat-
edly.[84] Nothing, of course, was farther from Goethe's mind than the explication of
elective attraction in terms of a rigorous theory. What he tried to find and explain in

this striking item of the new scientific theories was its moral force. In an advertisement that Goethe wrote for his novel he said of himself:

> It seems that his continued work in physics made the author choose this strange title. He may have noticed that often in natural science ethical similes are used to bring something nearer that is remote from the region of human knowledge, and so, presumably, he may have wanted to trace the parlance of a chemical simile back to its spiritual origin, all the more so since there is after all just one nature. . . .[85]

Goethe, then, seemed to think of his novel as a literary attempt to reunite the spiritual and the material, the moral and the physical. In the event, both Kant and Goethe were wrong about chemistry. *Pace* Kant, it became a rigorous quantitative science; *pace* Goethe, the elective affinity of elements need not and cannot be reduced to a spiritual origin but is explained in terms of the electropositive or electronegative structure of their electron shells.

A great work of art allows for endless interpretation. Goethe himself directs us to what is at least one important theme in the *Elective Affinities*—the force of contingency and presence. In his advertisement, he abruptly turned from the themes of science and coherence to the unpredictable and uncontrollable force of contingency. The quotation above continues as follows:

> since there is after all just one nature and even the realm of serene rational freedom is irresistibly pervaded by traces of turbid, passionate necessity that can only be erased fully by a higher hand and perhaps not even in this life.[86]

The necessity that concerns Goethe is not the formal necessity of logic or the nomic necessity of science but that of contingency.

The *Elective Affinities* takes place on a baronial estate, and the efforts of the four protagonists, of the men in particular, is directed toward gaining a lucid and explicit grasp of the estate and to giving it a more commodious and pleasing shape. They turn to chemistry because Charlotte, Baron Eduard's wife, is worried about the lead glaze on their pottery and verdigris on their copper containers.[87] When Eduard, Charlotte, and the captain (later promoted to major) apply the chemical puzzle of crosswise affinity to themselves, they obviously identify the first pair, AB with Charlotte and Eduard, and the second, CD, with the captain and with Charlotte's yet to arrive niece, Ottilie. But when it comes to the separations and affinities, Eduard assumes the harmless and amusing association of the two men and the two women. Their pleasant and constructive life, however, is upset by unforeseen and wrenching events. Charlotte and the captain are drawn to each other, Eduard and Ottilie fall hopelessly in love, Charlotte and Eduard have an unexpected offspring, Ottilie causes the infant boy to drown, and when finally, the way to pairwise dividing and crosswise uniting seems open, Ottilie renounces her happiness.

While contingency heightens presence, lawfulness diminishes it, theoretically by reducing full-bodied things to instances of scientific laws and practically by furnishing the explanations that make possible the technological control and reduction of the contingencies of hunger, cold, illness, and untimely death. Contingencies come in many sizes and shapes. The large and captivating incidents and accidents present themselves with commanding force. Once present, there was no preventing them through planning, and there is no escaping them now through a technological fix. If the event is positive, we are willing to accept it as sheer good fortune or just deserts. If it is negative, we tend to react with uncomprehending anger and the indignant protest: "I don't need that." Goethe's contemporaries too were brought up short by the darkness of contingency. The "serene rational freedom" of the Enlightenment had already become the ruling cultural assumption.[88]

If scientific explanation and rational planning are to no avail in the face of out-rageous fortune, what attitude is appropriate? It is exemplified by Charlotte, who from the start has respect for contingency and shows fortitude in the face of misery. When the captain suggests that the notion of *elective* affinity implies preference and a choice, Charlotte replies:

> Forgive me as I forgive the scientist; but I would not at all think of a choice here, much rather of natural necessity, and not even of that, for in the end it is perhaps just a matter of opportunity; opportunity makes relationships, just as it makes thieves.[89]

So far, however, Charlotte's attitude seems to point forward more to the existential-ist self-assertion in the face of absurdity than to the romantic recovery of continu-ity with nature and tradition. The treatment of nature in the *Elective Affinities* is a warning rather than an exemplar. The hills, creeks, ponds, rocks, and trees of the baronial estate are endlessly subjected to planning and improvement without grace and repose ever settling on it all. The recovery of tradition is tellingly triggered by an "incident" (*Vorfall*) concerning a dispute about an endowment for the cemetery near the castle.[90] The incident calls attention to the church inside the cemetery and finally leads to the discovery of a chapel, attached to the side of the church. Church and chapel are evidently in the Gothic style that had seemed barbarous and negligible to Enlightenment tastes. Charlotte, Ottilie, and a young architect now set about to restore and embellish the church and particularly the chapel.

One's heritage is a contingency too. You do not get to choose it from a catalog. Goethe stresses the point: "This church had been standing there for several cen-turies in the German style and art, erected in substantial proportions and happily adorned."[91] The concern of the three is to make that heritage present again—a formi-dable task. In the conversation about the disputed cemetery endowment, Charlotte reminds her listeners of "how hard it is to honor the present properly."[92] Goethe described the convergence of past and present that occurred in the church by saying that the church "grew, so to speak, toward the past."[93] Goethe further described the past that comes to be present in the ancient drawings that the architect shows Char-lotte and Ottilie and that serve as models for the figures that the architect and Ottilie are to paint on the vaults of the chapel. Of the archaic character of these drawings Goethe says: "…how engaging did the viewers find it! From all of these figures, pur-est presence was shining forth."[94]

The chapel becomes the place of rest and grace. Ottilie and Eduard were buried there. The novel ends with these sentences:

> Thus, the lovers are resting next to one another. Peace dwells over their place; serene, familiar images of angels are looking down on them, and what a pleasant moment it will be when some time in the future they will awake together.[95]

When joining Goethe's novel with present knowledge, we see the chemical and human elections and affinities separate from one another. The former are now scientifically and lawfully intelligible, the latter, by contrast rather than springing from a common root with science, are all the more commanding in their contingency and presence. Commanding presence is a large and varied theme in Goethe's writings and is often in tension with the theme of restless striving—a tension not always resolved though it is decided in favor of restful presence in the *Elective Affinities*. Presence and immediacy can be found at three different levels in the novel. Immediate presence in the broadest sense is simply part of Goethe's genius as a writer (and perhaps his limit as a philosopher and scientist as well). What Goethe always does is show rather than tell, and especially so in the *Elective Affinities* as Benno von Wiese has noted: "everything is presence, image, and figure; nothing is merely thought."[96]

Within this widest region of presence, a stronger sort of presence belongs to those archetypal entities that in an exemplary way unite the universal and the particular. Carl Friedrich von Weizsäcker accurately observed that the exemplar (*Gestalt*) rather than the law (*Gesetz*) is at the center of Goethe's conception of science.[97] Some scholars see in it a scientific ontology in its own right.[98] Weizsäcker thinks it hearkens back to a Platonic ontology of ideal forms.[99] Both of these claims seem mistaken to me.

As for Goethe's putative Platonism, it would conflict with Goethe's delight in the concrete presence of things. Spinoza was the early philosophical influence on Goethe, and from him he took the notion that tangible reality is entirely the presence of the spiritual. To this Goethe added his own profound pleasure in shaping reality. It is what he practiced as the master of his household, as director of the Weimar theater, and as a minister of the Duchy of Saxe-Weimar-Eisenach, and it is what the old Faust looks back to with satisfaction.[100]

Though Aristotle was a minor figure in Goethe's pantheon, the fourfold Aristotelian causality serves best to explicate Goethe's view of how humans should shape reality. The transformation of the baronial estate was an exemplar of Aristotelian craft with Eduard's friend, the captain, to be the efficient cause of the fusion of the material and formal causes. As Eduard explains to Charlotte:

> Country folk have the right knowledge; however, the information they provide is confused and not candid. The experts from the city and the academies are, to be sure, clear and consistent, but they lack immediate insight into things. My friend promises to furnish both.[101]

Alas, no final cause of rightful and persuasive authority was available to Goethe, nor was the hierarchy of being that would lend Aristotelian forms their place in a cosmic order, nor, finally, did Spinoza's plenitude and perfection of God or nature have the kind of forceful presence that would constitute a background (inconsistent, to be sure, from Spinoza's point of view) for the ordering of reality.

"Goethean science" has been captured by outsiders and cranks. Though Goethe himself has aided and abetted his scientific acolytes, he has left few traces in serious contemporary science. More to the point here, the world of Aristotelian exemplars and Spinozist necessity is overwhelmed in the *Elective Affinities* by a yet more formidable presence, implacable first and finally consoling. Goethe's conception of reality is, of course, for the most part implicit in his literary works of art. In the *Elective Affinities* it is richly presented and morally eloquent and yet imperiled by doubts and deficiencies. The most elaborate account of the good life in the *Elective Affinities* is the description of the calm before the catastrophes and the respite between the catastrophes and the final denouement. It is a highly educated life, musical, enterprising, and companionable, enacted in a prosperous, graceful, and engaging setting. But it is a brittle world, finally shattered by contingency; and though rest and grace prevail in the end, the final word is a promise with little hope today of durability and prospect of detailed realization.

The rigorous conception of reality in Newtonian physics was for Goethe an incomprehensible iron cage. The distant rumble of the rising Industrial Revolution that was to reshape terrestrial reality from the ground up Goethe observed with helpless anxiety. And the deeply engrained if benign paternalism of the *Elective Affinities* was insensible to the democratic impulses that were about to reform social reality. Goethe's enduring bequest, however, is the appreciation of the splendor and presence of reality.

THE RECONCILIATION OF FREEDOM
AND LAWFULNESS—SCHELLING

It was Friedrich Wilhelm von Schelling (1775–1854) who recognized that Kantian lawfulness and Goethean presence had to be reconciled with one another. While still in his teens, he studied Kant at the Tübinger Stift, the famous college where the spiritual and administrative elite of the Duchy of Württemberg was trained and where, as it happened, many a poet and philosopher was educated as well. Schelling took courses at the Stift in mathematics and in theoretical and experimental physics. After the conclusion of his studies and a year of teaching as a private tutor, Schelling went to the University of Leipzig where he attended lectures in physics, chemistry, and medicine.[102] Thus, unlike Goethe and much like Kant, Schelling took the physical sciences seriously. He was well-read in the scientific literature of his day, and he was determined to take on the task of coming to philosophical terms with rigorous quantitative science.

Like Goethe and unlike Kant, Schelling was taken with the datum of all data, with the unsurpassable givenness of nature.[103] This awakening to natural reality was an original and surprising event in young Schelling's life. His philosophical curiosity and self-confidence had been stimulated by a meeting with Fichte in 1794 while Schelling was still a student at the Tübinger Stift. But rather than follow Fichte in tracing all there is to the sole and original givenness of the ego, Schelling pleaded for the equally primordial datum of nature.[104] The results of this remarkable turn of thought were Schelling's *Ideas for a Philosophy of Nature* of 1797 (when Schelling was twenty-two) and of *On the World Soul,* published the following year.[105] Goethe recognized a soulmate in Schelling—sometime between 1798 and 1802 he responded with his poem "World Soul."[106] They met in 1798; the result was a lifelong friendship.

What Schelling had in common with both Kant and Goethe was the surrender of astronomical cosmology to physics. To Schelling too, the large-scale scientific structure of the universe, a few comments aside, seemed morally barren. He also shared with Kant and Goethe the particular point of departure in the search for a moral cosmology: The critique and the transcendence of atomist and mechanical Newtonian natural philosophy. There were differences of emphasis, to be sure. Schelling was much more pointed than Kant in his criticisms of Newtonian science, and contrary to Goethe, Schelling was concerned primarily with Newton's physics rather than Newton's optics.[107] Schelling is notorious for his variable and impressionist approaches to philosophy.[108] Thus, unsurprisingly, he did not work out a resolution of the Kantian-Goethean conflict in his seminal books, but rather furnished a variety of instructive explorations. Here again the notion of physical lawfulness provides a thread we can follow in tracing Schelling's search for a moral cosmology.

One way of recovering a moral cosmology that Schelling pursued was to limit the scope and significance of the scientific theories that threatened to reveal a merely mechanical universe. It is a path that has remained dear to humanists and social scientists. They counter the claim that physics gives us a uniquely penetrating and accurate account of reality with the claim that scientific theories are in the first instance not about reality at all, but are merely convenient fictions or instruments for the prediction and control of the reality we are all familiar with. Schelling gave a clear statement of what has come to be called scientific instrumentalism in his *Ideas for a Philosophy of Nature*:

> It is precisely this that in large part constitutes the task of a philosophical theory of nature, viz., to determine the admissibility as well as the limits of those fictions in physics that are simply necessary for further progress of investigation and observation and that impede scientific progress only if we intend to employ them beyond their limit.[109]

There is evidently a Kantian impetus in Schelling's realization that the physical lawfulness of reality needs to be acknowledged somehow; and, *pace* Goethe, he agreed with Kant that this lawfulness has a mathematical structure. In fact, he went beyond Kant in holding out for the inclusion of chemistry among the properly scientific theories:

> For, if a theory of nature is a *science of nature* only to the extent that mathematics can be employed in it, then a system of chemistry that admittedly rests on false presuppositions,

but is in a position, on the strength of those presuppositions, to render this experimental doctrine mathematically, is preferable to one that may have the merit of resting on true principles, but, those principles notwithstanding, must give up on scientific precision (on the mathematical construction of the respective phenomena).[110]

Yet unlike Kant, Schelling held that what today we call theoretical objects fail to *explain* phenomena. Speaking of "magnetic matter" as a theoretical object, Schelling says: "To assume the latter is fine as long as it is only taken as a (*scientific*) *fiction* that is used as a basis for *experiments* and *observations* (as a *heuristic device*), but not for *explanations* and *hypotheses* (as a *principle*)."[111]

What fell outside the scope of the fictional objects and laws of physics, according to Schelling, were the phenomena of force and organism.[112] To mark out a realm of agency and life was to recover the contours of a moral cosmology. Schelling knew of course that force was prominent in Newtonian physics as universal attraction and as the product of mass and acceleration in the second law of motion.[113] Regarding the force of attraction, Schelling, following Kant, tried to show its inadequacy by arguing that attraction without repulsion is unintelligible. In this line of attack, he felt much encouraged by Newton's own misgivings about a force that acted instantaneously at a distance. In the general Scholium to the *Principia*, Newton acknowledged that he had "not yet assigned a cause to gravity," and, having summarized the effects of gravity in the solar system, he conceded that he had "not as yet been able to deduce from phenomena the reason for these properties of gravity." Yet he pronounced himself satisfied "that gravity really exists and acts according to the laws that we have set forth and is sufficient to explain all the motions of the heavenly bodies and of our sea."[114]

Schelling, however, denied that, absent a causal account, it is legitimate to claim that an explanation has been forthcoming:

> If therefore Newton was really doubtful, as he says in several places (disregarding others where he explicitly claims the opposite) what "the actual cause of attraction" is supposed to be, whether it is not perhaps actuated by an impulse or in some other way, unknown to us, then the use he made of that principle in the construction of a world system, was in fact mere semblance of a use; or rather the force of attraction itself was for him a scientific fiction that he used merely to reduce the phenomenon as such to laws without thereby intending to explain it.[115]

What Schelling found missing in Newtonian mechanics were "spontaneously moving forces."[116] At times he hoped to locate them in those fields where mechanics had so far failed to provide coherent and comprehensive explanations, viz., in chemistry and electricity.[117] Most remarkably, Schelling thought that chemistry as a science would reveal how living matter sought "to step out of its equilibrium and to surrender to the free play of forces."[118] The marks of life and agency for Schelling were freedom and contingency while mechanics was the science of necessity.[119] What remained unexplained was how chemistry as the mathematical discipline that Schelling was looking forward to would consist with chemistry as the science of contingency (or of emergence). Analogously it was unclear how the recognition of repulsion as the supposedly necessary complement to attraction would be compatible with and perhaps complementary to attraction and gravity. Just as Kant finally had to all but emasculate his favored metaphysical-dynamical hypothesis and salute the philosophically flawed mathematical-mechanical version, so Schelling damned the repulsive force with faint praise and conceded explanatory power to the attractive force of gravity:

> Hence, we cannot use the repulsive force in its application any further than to make basically comprehensible how a material world is possible. As soon, however, as we try to explain how a certain system of the world is possible, the repulsive force does not take us even one step ahead.
>
> The structure of the heavens and the motions of the heavenly bodies we can explain only and solely by virtue of the laws of universal attraction.[120]

Still, concern with force remained one of the hallmarks of Schelling's philosophy of nature, and on occasion he called his approach "dynamical philosophy" to distinguish it from the mechanical philosophy of the Newtonians.[121]

The *Ideas for a Philosophy of Nature* of 1797 was concerned with physics, chemistry, magnetism, and electricity. The study of organic nature was projected for a later work.[122] Although Schelling denied that his *On the World Soul* of 1798 was the sequel to the *Ideas*, it clearly centers on the notion of organism.[123] In fact the full title of the book reads: *On the World Soul: A Hypothesis of Higher Physics Toward the Explanation of General Organism.* Schelling was searching for the seat and origin of force and life and at one point placed it in the infinite spaces of the universe. He declared that "in these regions actually lies the source of those inexhaustible forces that in individual matter spread in all directions and that compel and sustain movement and life on the solid cosmic bodies."[124] This, he adds, is where we find the "fullness of force that, being generated ever afresh in the depths of the universe, pours forth in individual streams from the center toward the periphery of the world system."[125] Remarkably, some of Schelling's demands and proposals were realized by Einstein's theory of general relativity.

There was, no doubt, a grand vision of a moral cosmology in Schelling's declarations. But it confronted him with two problems—(1) how to make this vision of a life-giving cosmos articulate and fruitful for everyday life and (2) how to square this vision with the part of contemporary cosmology that had assumed hegemony as Newtonian physics. Schelling's answer to the first question was indirect and was given some ten years later. The answer to the second question was characteristic of the openness and insight of young Schelling. He stated the problem as a dilemma. Speaking of the failed efforts of "most natural scientists" to appreciate "the meaning of the problem of the origin of organized bodies," he said:

> If some of them posit a special *life force* that, as a magic power, suspends all effects of natural laws in a living being, they thereby suspend *a priori* any possibility of explaining organization physically.

If, to the contrary, others explain the origin of all organization on the basis of dead chemical forces, they thereby suspend any freedom of nature in formation and organization.[126]

In a formulation reminiscent of Kant's effort to reconcile the freedom of imagination with the lawfulness of understanding, Schelling sketched the path between the
horns of the dilemma as follows: "Nature, in its blind lawfulness, must be free, and
conversely must be lawful in its full freedom; in this union alone lies the concept of
organization."[127] He hoped that this union could be accomplished through the notion of a "formative drive." But in the end, he realized that the formative drive had
no more explanatory power than the repulsive force. For purposes of explaining the
union of freedom and lawfulness, he realized, the formative drive is no more than
"a barrier for investigative reason or the cushion of an obscure quality on which to
get reason to rest."[128] The bulk of *On the World Soul*, as well as that of *Ideas*, is devoted, however, not to such revealing overviews, but rather to detailed discussions of
contemporary scientific problems and to attempts at aiding their solutions through
theoretical considerations. Some of Schelling's extended conjectures strike us as bi-
zarre today.[129] Yet Schelling was willing to submit to the discipline of experience and
often called for experiments that would decide the fate of his conjectures.[130]

Schelling's two early books won him Goethe's admiration and friendship and, as a
consequence, a professorial appointment and regular contact with Goethe in Jena.[131]
How this interchange influenced the intellectual careers of the young philosopher
and the mature poet is controversial and difficult to adjudicate. In his Jena years, at
any rate, Schelling's comments on the natural sciences became more impatient and
peremptory.[132] A dialectical monism of mind and matter began to eclipse and cover
up the dilemma Schelling had uncovered in the *World Soul*. Eventually, art (and later
religion) took the place of nature as the great datum philosophy had to acknowledge
and understand. In the introduction to his lecture *Philosophy of Art*, delivered in
the winter semester of 1802–1803, Schelling marked his change of focus as follows:

If we feel irresistibly driven to behold the inner essence of nature and to uncover that
fruitful source that, with eternal uniformity and lawfulness, pours out of itself so many
grand phenomena, how much more must it be of interest to us to penetrate the organism of art wherein, out of absolute freedom, supreme unity and lawfulness constitute
themselves which allow us to recognize the wonders of our own spirit much more immediately than nature.[133]

Evidently, the reconciliation of freedom and lawfulness remained Schelling's concern, but he now thought that its actuality could be detected both in nature and in
art. In his *Further Presentations from the System of Philosophy*, also of 1802, Schelling
specified the difference and consonance of these reconciliations and claimed cosmic
significance for the reconciliation of the reconciliations.

The universe, taken as an absolute, is articulated as the most perfect organic being and
as the most perfect work of art—for reason, which recognizes it in the absolute, it is

so *qua* absolute truth, for imagination, which represents it, it is so *qua* absolute beauty. Each of these simply expresses the same unity from different sides, and both coincide in the absolute point of balance in whose recognition lies at once the beginning and the goal of all knowledge.[134]

This passage suggests a complementary relationship between imagination and reason and, by extrapolation, of art and science. Robert J. Richards in an insightful and helpful essay has argued that "the aesthetic-epistemic principle of the complementarity of the poetic and scientific conceptions of nature" was "a fundamental organizing conception in the philosophy of the early Romantics."[135] Complementarity, he further urges, is also the resolution Schelling helped to achieve of the tension between Kant and Goethe and between lawfulness and freedom. Although I have developed these tensions in ways that differ from Richards's argument, I accept and am indebted to his principal claim—complementarity is the resolution to, as I put it above, the dilemma of lawfulness and contingency. I would like to offer, however, a modification or extension of Richards's complementarity that bears significantly on the search for a moral cosmology.

There are two connected features of complementarity as Richards understands it that need clarification. Richards finds his notion of complementarity both in Schelling at the beginning of the nineteenth century and in Helmholtz at the close of the century. Schelling, Richards says, "theoretically demonstrated that scientific understanding and artistic intuition did not play out in opposition to one another, as Goethe once thought, but that they reflected complementary modes of penetrating nature's underlying laws."[136] Helmholtz's conception of complementarity, drawn from Goethean sources, Richards summarizes thus: "Both aesthetic intuition and scientific comprehension drove down to the type, to the underlying force that gave form to the surface of things."[137]

The first problem lies in the putative point of convergence of science and art. The first quotation from Richards suggests that science and art converge on "underlying laws." These are characterized by Goethe as "an unknown law-like something" and as the "secret laws of nature," and by Richards (following Schiller) as "ineffable rules," and (following Schelling) as "inarticulable laws," and as "nature's concealed laws."[138] Taken at face value, these laws are the notorious wheels that turn without turning anything else. There can be no substantive disagreements about them because they are unavailable for appeal or proof. They cannot link causes and effects because we do not know their causal structure. They cannot be the laws that concern the natural sciences because those laws are known and articulate, and where they are not, science does not rest until they are discovered and spelled out.

Hence there can be no convergence of art and science in underlying laws. There can be a kind of complementarity, and here we come to the second problem of Richards's argument. This complementarity might be called *alternative complementarity*, the sort that is familiar from the Copenhagen interpretation of quantum theory, where our inability to comprehend the particle model and the wave model of, say, the electron in one single model, has led to the decision to baptize the problem "complementarity" and be done with it.

If alternative complementarity is the final word, the romantic search for a coherent and morally promising cosmology is no closer to its goal than was Schelling before he met Goethe. There is, however, a second kind of complementarity, the more usual sort actually, the one we may call *interdependent complementarity*. Two things are complementary in that sense if they are distinct but cannot be thought or exist apart from one another. A simple example is the relation of matter and form. That perhaps we should not take *law*, as Richards, following Goethe, Schiller, and Schelling, uses it, at face value, i.e., in the sense of an explicit and informative assertion that carries nomic necessity and a universal quantifier, appears from Richards's summary of Helmholtz's view where it is the "type" and the "underlying force," and not necessarily a law, that art and science converge on.[139] Goethe too uses law in the sense of "exemplar" and in the wider sense of cultural or artistic constraint.[140]

The interdependent complement of a law of nature is the instance that constrains the law to yield the description and explanation of a state of affairs or an event. Newton's laws of motion merely outline a possibility space. They describe an actual world when we insert the values of, e.g., the solar system in place of the variables for mass, acceleration, distance, etc. Among the greatest of instantiations are works of art. They are the most eminent complements to laws. They are instances of high contingency—unpredictable and unprocurable and, in that sense, free. So are the nuisances of life and the results of throwing dice. But these are part of the low contingency of everyday reality. Works of art rise above and lend orientation to the plains of normalcy.[141]

Between these landmarks and the laws of science are intermediate instantiations, regularities that possess neither the universal necessity of laws nor the commanding presence of art works. Among them are the natural kinds of everyday life, the species of biology, and the genres of literature. Goethe's preoccupation with regularities was a distraction, it seems to me, and led him into a hopeless competition with scientific lawfulness. His great accomplishment was the disclosure of presence. The solution to Schelling's problem is the interdependent complementarity of Kantian lawfulness and Goethean presence. Before I suggest how this romantic bequest can bear moral fruit today, I have to remark briefly on how Kant, Goethe, and Schelling instruct us as regards the second term of a moral cosmology—the cultural receptivity, or the lack of it, for cosmic significance.

Kant was most attuned to the emancipatory aspirations of his age though he did not escape the ethnic and masculine prejudices of the times. His hope for an era of equality, dignity, and self-determination (the three aspects of the moral law) was tempered by his dark view of human goodness, and it was largely untouched by the rising Industrial Revolution.[142] Goethe contemplated the rise of machines with dismay and put little store by the ability of the bourgeoisie to deal intelligently with the challenges of the age. In the "Prologue in Heaven" he has Mephisto explain why there was little hope that people might join in the cosmic contemplation of the archangels. Mephisto likens humanity to a grasshopper:

that ever flies and, flying, leaps
and in the grass still sings its same old song;
if only it would always lie there in the grass!
It sticks its nose in every nuisance.[143]

Schelling generally paid little attention to everyday culture and people, and in a lecture, where he did so, in his "On the Essence of German Science" of 1807, he justified this neglect by claiming that the deplorable state of German culture was beyond philosophical illumination and redemption.[144] More important, among the causes of this calamity he mentioned "mechanism," an implicit extension of the mechanical ontology of Newton to the cultural and political sphere. "Complete mechanization of all talents, all history and institutions," Schelling said, "is here the highest goal."[145] Not that the state achieves wholeness in this way, but the reason for this failure

is always sought in the imperfection of mechanism; new wheels are installed that need still other wheels for their regulation, and so on ad infinitum; what remains ever equally distant is the mechanical *perpetuum mobile* whose invention is reserved entirely for the organic art of nature and humanity.[146]

LAWFULNESS AND CONTINGENCY

The romantic bequest for today's task of finding a moral cosmology tells us that we need to discover the moral force of the complementarity of lawfulness and contingency for a technological society. First of all, then we have to follow Kant and Schelling in taking contemporary scientific cosmology seriously. It presents great obstacles and opportunities. The obstacles lie in the astoundingly unsettled state of contemporary astrophysics. First, more than 90 percent of the universe is unaccounted for. We know of these missing parts only because of their gravitational force (dark matter) or their accelerating force in driving the expansion of the universe (dark energy). Second, the two theoretical pillars of contemporary physics, quantum theory and relativity theory, are inconsistent with one another. And third, there is at least a possibility that ours is not *the* universe but only a phase or a part of a multiverse, a system of many, perhaps infinitely many and perhaps fundamentally different, universes.[147]

But new vistas and opportunities have opened up as well. Hubble's discovery of the expanding universe in the late 1920s led by way of retrospection to the discovery of the Big Bang, and this in turn reunited astronomy and physics to give us astrophysics. There is, moreover, a solid majority view in astrophysics that we should proceed on the assumption that there is just one universe, ours; and there is a somewhat less solid majority view that a unified theory, a "final theory," of relativity theory and quantum theory will be reached and that string theory is currently the most promising contender.

There are now the beginnings of moral cosmologies that depart from some outline of the final theory. There is Frank Tipler's provocative, *The Physics of Immortality*.[148] Eric Chaisson disavows any moral or teleological intentions, and yet exhibits an enthusiasm that seems to gesture at a moral cosmology.[149] Brian Swimme and Thomas Berry have told a creditable story of the cosmos that begins broadly and then zeroes in on our planet and our environmental obligations.[150] The most magnificent moral cosmology is Robert Pack's *Before It Vanishes*, a cycle of thirty-one poems that ponder and respond to Heinz R. Pagels's *The Cosmic Code*.[151] It unites scientific lawfulness with the poetic presence of nature and culture. Philosophers need to catch up with Pack, and just to point in the direction of that task let me try to show how astrophysics allows us to trace a remarkable path from lawfulness to contingency and to the threshold of a moral cosmology.

The notion of force that so occupied Schelling has divided into four kinds in the twentieth century, and the task since has been to discover their original unity and the history of their unfolding. At the beginning of the process stands the symmetry of the cosmos. The early universe was the most symmetrical possible. In mathematics and physics, symmetry is a principle of invariance that connects with what we ordinarily mean by symmetry, but then also greatly exceeds the ordinary sense of symmetry. It makes common sense to say that a daisy has a greater symmetry than a lady's slipper. The latter looks the same from left and right. The daisy, however, looks the same no matter how you turn it on its vertical axis. A sphere is more symmetrical yet because, unlike a daisy, it looks the same however you turn it on any of its infinitely many axes. These are spatial symmetries, but there are also symmetries of time, illustrated, e.g., by laws that hold today as well as tomorrow and in fact at any time whatever. There are more recondite symmetries, and altogether there are a half a dozen of them.[152]

In the early universe there was no distinction between gravity, electromagnetism, the strong force, and the weak force. As the temperature fell, this symmetry was broken. First the gravitational force crystallized out from the other three, then the strong force, and finally the weak and electromagnetic forces separated from one another. The Holy Grail of physics is a theory that would unite all four forces. A step toward unification was the fusing of the weak and electromagnetic forces (the latter force itself representing a unification, achieved by James Clerk Maxwell in 1865). Steven Weinberg provided a unification of the weak and the electromagnetic force in 1967.[153] They govern, as far as we know, today's entire universe. But there is also a local breaking of symmetries. The geology of Earth is notably asymmetrical so that, when something natural approaches symmetry, it becomes noteworthy for that reason, a mountain peak that approximates the rotational symmetry of a pyramid (like Haystack Butte in Montana) or the bilateral symmetry of a stone arch (like the Eye of the Needle above the Missouri in Montana, now destroyed). Many conifers and flower heads are instances of rotational symmetry. The full moon and the ripples of a stone thrown in a calm pond are fleeting examples of perfectly round symmetry. Most remarkable is the bilateral symmetry of most animals, and it grows in force from fish to lions and finally to humans who in their erect posture display bilateral symmetry most fully. This is the least kind of symmetry. It is entailed by all others and entails no other. Hence when it comes to symmetry, the Earth, its inhabitants, and humans particularly, occupy a singular position.

Part of a cosmic symmetry is the homogeneity and isotropy of the universe. It is homogeneous in that it looks the same from every location and isotropic in that it looks the same in all directions. This is so for the universe at large. At the scale of galaxies there are differences of shape and distance. The solar system is still less homogeneous and isotropic. Every human being, finally, embodies a strongly oriented space with the significantly different up-and-down, front-and-back, and left-and-right orientation that is reflected in the hundreds of metaphors we use all the time.[154] Likewise the natural and cultural space humans inhabit is oriented around landmarks.

The breaking of symmetries leads us from lawfulness to contingency, from causal links to unforethinkable presence. We owe to technology the instruments that have enabled us to discover nuclear physics through accelerators and to discover the astronomy of space and time through terrestrial and satellite telescopes. But technology as a culture has also made us unaware of the starry heavens through light pollution and unconcerned about science and ethics through the distractions of consumption. Can ethics lead us through consumption to a life of engagement in the things and practices that matter in the reality of daily life?

NOTES

1. Caleb Everett, *Numbers and the Making of Us* (Cambridge, MA: Harvard University Press, 2017), pp. 36–37.

2. George Bird Grinnell, *Blackfeet Indian Stories* (Helena, MT: Riverbend, 2005 [1913]), pp. 97–103.

3. James Welch, *Fools Crow* (New York: Viking Penguin, 1986), pp. 108–17.

4. Charles Taylor, *A Secular Age* (Cambridge, MA: Harvard University Press, 2007), pp. 146–58.

5. See my "The Headaches and Pleasures of General Education," *The Montana Professor*, volume 13, pp. 10–15.

6. Joel R. Primack and Nancy Ellen Abrams, *The View from the Center of the Universe* (New York: Riverhead Books, 2006), pp. 3–6.

7. Galileo Galilei, "The Assayer," in *The Controversy on the Comets of 1618* (Philadelphia: University of Pennsylvania Press, 1960 [1623], p. 184.

8. Brian Greene, *The Elegant Universe* (New York: W. W. Norton, 1999), p. 139.

9. George Lakoff and Mark Johnson, *Metaphors We Live By* (Chicago: University of Chicago Press, 1980).

10. Wolf-Dieter Gudopp von Behm, *Thales und die Folgen: Vom Werden des philosophischen Gedankens* (Würzburg, Germany: Königshausen und Neumann, 2015).

11. Immanuel Kant, *Kritik der reinen Vernunft*, 2nd ed., ed. Raymund Schmidt (Hamburg, Felix Meiner, 1956 [https://guava.physics.uiuc.edu/~nigel/courses/569/Essays_Fall2007/files/xianhao_xin.pdf87]), pp. 448–69 [pp. 448–89 in the Prussian Academy edition].

12. Aristotle, *On the Heavens*, ed. W. K. C. Guthrie (Cambridge, MA: Harvard University Press, 2000 [1939]).

13. Aristotle, *On the Soul*, ed. W. S. Hett (Cambridge, MA: Harvard University Press, 1957).

14. Aristotle, *Nicomachean Ethics*, ed. H. Rackham (Cambridge, MA: Harvard University Press, 1962 [1934]), pp. 62–67 (1102a26–1103a10 in the Bekker pagination).

15. Aristotle, *On the Soul*, p. 178 (431b21–22 in the Bekker pagination).

16. Aristotle, *Metaphysics*, ed. Hugh Tredennick (Cambridge, MA: Harvard University Press, 1958 [1935]), pp. 122–75 (1069a1–1076a4 in the Bekker pagination).

17. Thomas Aquinas, *Summa Theologica*, 16th ed., ed. Nicolaus Sylvius Bouillard and C. J. Drioux (Paris: Bloud and Barral, 1867–1874), vol. 1, pp. 57–60 (Pars Prima, Quaestio X).

18. Nicolaus Copernicus, "On the Revolutions of the Heavenly Spheres," *Theories of the Universe*, ed. Milton K. Munitz (New York: The Free Press, 1957), pp. 150–51.

19. Copernicus, p. 151.

20. Copernicus, p. 169.

21. Ibid.

22. Steven Weinberg, *Dreams of a Final Theory* (New York: Pantheon Books, 1992), p. 257.

23. Galileo Galilei, *Dialogue Concerning the Two Chief World Systems*, tr. Stillman Drake, 2nd ed. (Berkeley: University of California Press, 1967 [1629]), pp. 23–28.

24. Galileo, *Assayer*, pp. 183–84; *Dialogue*, pp. 230 and 233.

25. Galileo, *Dialogue*, p. 341.

26. Ibid.

27. Blaise Pascal, "Le Mémorial,"*Pensées*, ed. Victor Giraud (Paris: Rombaldi, 1935), pp. 1–15.

28. Martin Heidegger, *Identität und Differenz* (Pfullingen: Neske, 1957), p. 70 (my translation).

29. Johannes Kepler, "Praefatio Antiqua ad Lectorem" in *Mysterium Cosmographicum*, 1706 (Paris: Erasmus Kempfer and Godefridus Tampachius, 1721), p. 10. This is from a reproduction of the original (London: Dalton House, 2018).

30. Kepler, *Astronomia Nova*, translated and edited by William Donahue (Santa Fe, NM: Green Lion Press, 2015 [1609], chapters 48 and 49, pp. 230–369 [230–39].

31. Isaac Newton, *The Principia: Mathematical Principles of Natural Philosophy*, translated and edited by Bernard Cohen and Anne Whitman (Berkeley: University of California Press, 1999).

32. Newton, *Principia*, pp. 416–17.

33. Ernest Nagel, *The Structure of Science* (New York: Harcourt, Brace & World, 1961), p. 154.

34. Nagel, p. 155.

35. Newton, *Principia*, pp. 382–83.

36. Newton, *Principia*, p. 943.

37. Newton, Isaac, *Opticks* (Amherst, NY: Prometheus Books, 2003), pp. 368–69.

38. Newton, *Opticks*, pp. 368–70.

39. Newton, *Principia*, p. 940.

40. Newton, *Principia*, p. 942.

41. Newton, *Principia*, p. 943.

42. Kant, Immanuel, *Universal Natural History and Theory of the Heavens,* in *Theories of the Universe: From Babylonian Myth to Modern Science,* Milton K. Munitz, ed. (New York: The Free Press, 1957), p. 237.

43. Kant, *Kritik der praktischen Vernunft,* ed. Karl Vorländer (Hamburg: Felix Meiner, 1963 [1787]), p. 186 (my translation).

44. Gardner, *Relativity Simply Explained* (Mineola, NY: Dover Publications, 1997 [1962]), p. vii.

45. Charles R. Sunderlee, "A Thrilling event on the Yellowstone," *The Helena Daily Herald,* Wednesday, May 18, 1870, p. 1.

46. Charles M. Skinner, *Myths and Legends of Our Own Land* (no place [Canton, OH]: Pinnacle Press, no date), pp. 347–48. The book was originally published in 1896 by J.B. Lippincott in Philadelphia.

53 Ella E. Clark, *Indian Legends of the Northern Rockies* (Norman: University of Oklahoma Press, 1988 [1966]), pp. 323–324.

54 Peter Nabokov and Lawrence Loendorf, *Restoring a Presence: American Indians and Yellowstone National Park* (Norman: University of Oklahoma Press, 2004), pp. 23–24.

55 Lee H. Whittlesey, "Native Americans, the Earliest Interpreters: What is Known About Their Legends and Stories of Yellowstone Park and the Complexities of Interpreting Them," *The George Wright Forum,* volume 19, number 3 (2002), p. 47.

47. Johannes Hofer, *Dissertatio Medica de Nostalgia oder Heimwehe* (Basel, Switzerland: Jacob Bertschi, 1688).

48. Hofer, p.14. My translation.

49. Cisco Lassiter, "Relocation and Illness: The Plight of the Navajo," *Pathologies of the Modern Self,* ed. David Michael Levin (New York: New York University Press, 1987), pp. 221–30.

50. Welch, p. 111.

51. Edward S. Casey, "The World of Nostalgia," *Man and World,* vol. 20 (1987), pp. 361–84. See also Svetlana Boym, "Nostalgia and Its Discontents," *The Hedgehog Review,* vol. 9, no. 2 (Summer 2007), pp. 7–18; and Filiberto Fuentenebro and Carmen Valiente Ots, "Nostalgia: A Conceptual History, *History of Psychiatry,* vol. 25 (2014), pp. 404–11.

52. Robert McC. Netting, *Balancing on an Alp* (Cambridge: Cambridge University Press, 1981).

Helen Güdel, *Lieber Alex,* volume 1: *Von Menschen und Tieren im Walliser Bergdorf Törbel* (Zurich: Atlantis Verlag, 1991) *Lieber Alex,* volume 2: *Briefe aus dem Walliser Bergdorf Törbel* (Zurich: Atlantis Verlag, 1993); volume 3: *Briefe aus dem Walliser Bergdorf* (Berne, Switzerland: Zytglogge Verlag, 1995).

53. Helen Güdel, *Lieber Alex,* vol.1: *Von Menschen und Tieren im Walliser Bergdorf Törbel* (Zurich: Atlantis Verlag, 1991); *Lieber Alex,* vol. 2: *Briefe aus dem Walliser Bergdorf Törbel* (Zurich: Atlantis Verlag, 1993); *Lieber Alex,* vol 3: *Briefe aus dem Walliser Bergdorf Törbel* (Berne, Switzerland: Zytglogge Verlag, 1995).

54. Steven Hawking, *A Brief History of Time* (New York: Bantam Books, 1988), 175.

55. Immanuel Kant, *Kritik der praktischen Vernunft,* Akademie-Ausgabe (Berlin: Georg Reimer, 1913), vol. 5, p. 161.—All of the following translations from the German are mine.

56. *The Cambridge Dictionary of Philosophy,* ed. Robert Audi (Cambridge: Cambridge University Press, 1995).

57. Ernan McMullin, "Cosmology," *Routledge Encyclopedia of Philosophy*, ed. Edward Craig (London: Routledge, 1998), vol. 2, pp. 677–81.

58. "Philosophy of Cosmology." Stanford Encyclopedia of Philosophy, Stanford University, 26 Sept. 2017, https://plato.stanford.edu/entries/cosmology/.

59. Virgil, *Aeneid*, ed. G.P. Goold, tr. H. Rushton Fairclough (Cambridge, MA: Harvard University, 1994 [1916]), vols. 1 and 2.

60. "Thales," in *Fragmente der Vorsokratiker*, ed. and tr. Hermann Diels (Zurich: Weidmansche Buchhandhandlung, 1964 [1952]), vol. 1, pp. 67–81.

61. Aristotle, *The Physics*, ed. and tr. Philip H. Wicksteed and Francis M. Cornford (Cambridge, MA: Harvard University Press, 1970 [1929]), vol. 1 and 2.

62. Aristotle, *The Metaphysics*, ed. and tr. Hugh Tredennick (Cambridge, MA: Harvard University Press, 1961 [1933]), vols. 1 and 2; *On the Heavens*, ed. and tr. W. K. C. Guthrie (Cambridge, MA: Harvard University Press, 2000 [1939]; *The Nicomachean Ethics*, ed. and tr. H. Rackham (Cambridge, MA: Harvard University Press, 1962 [1926]).

63. Isaac Newton, *The Principia*, tr. I. Bernard Cohen and Ann Whitman (Berkeley: University of California Press, 1999).

64. Newton refers to Copernicus (1473–1543) in *The Principia* on pp. 803–4 and to Kepler (1571–1630) on pp. 800, 805, and 924.

65. René Descartes, *Discourse on Method*, tr. Laurence J. Lafleur (no place: The Library of Liberal Arts, 2nd ed., 1956 [1637]), pp. 15–16.

66. David Hume, *A Treatise of Human Nature* (Oxford: Clarendon Press, 1968 [1737]), p. 469.

67. Karl Marx and Frederick Engels, *The Communist Manifesto (1848)*, Part One.

68. Ibid.

69. Isaac Newton, *Philosophical Writings*, Andrew Janiak, ed., Cambridge University Press, 2004, p. 90.

70. Immanuel Kant, *Universal History and Theory of the Heavens or Essay on the Constitution and Mechanical Origin of the Entire Cosmic Edifice, Carried Out According to Newtonian Principles*, 1755, Part 3.

71. Immanuel Kant, *Universal Natural History and Theory of the Heavens, or an Attempt to Account for the Constitutional and Mechanical Origin of the Universe upon Newtonian Principles* (1755), in *Kant: Natural Science*, Eric Watkins, ed. Cambridge University Press, 2012, p. 222.

72. Immanuel Kant, *Metaphysical Foundations of Natural Science* (1786). Cambridge University Press, 2004, translated and edited by Michael Friedman.

73. Johann Wolfgang von Goethe, *Wilhelm Meisters Wanderjahre*, Hamburger Ausgabe (Munich: Beck, 1973), vol. 8, p. 429.

74. See Clemens Brentano's mordant "Der Philister" ("The Philistine") in Clemens Brentano, *Werke*, ed. Friedhelm Kemp (Munich: Hanser, 1963), vol. 2, pp. 959–1016 and Frederick C. Beiser, *Enlightenment, Revolution, and Romanticism* (Cambridge: Harvard University Press, 1992).

75. Goethe, *Wilhelm Meister*, pp. 429 and 430.

76. Beiser, pp. 225–27.

77. *Goethes Gespräche mit Eckermann*, ed. Franz Deibel (n.p.: Insel-Verlag, 1949), p. 124.

78. Johann Wolfgang von Goethe, *Faust*, Hamburger Ausgabe (Hamburg: Christian Wegner [1966]), vol. 3, pp. 16 (lines 243–70).

79. *Gespräche mit Eckermann*, p. 278. See also Goethe's "Erfahrung und Wissenschaft," Hamburger Ausgabe (Hamburg: Christian Wegner, 1971), vol. 13, pp. 23–25.

80. *Gespräche mit Eckermann*, p. 396.

81. Immanuel Kant, *Metaphysische Anfangsgründe der Naturwissenschaft*, Hartknoch, 1787.

82. An account of Goethe's long-standing interest in elective affinity can be found in Benno von Wiese's notes to *Die Wahlverwandtschaften*, Hamburger Ausgabe (Hamburg: Christian Wegner, 1968), vol. 6, pp. 680–84.

83. Goethe, *Wahlverwandtschaften*, p. 276.

84. Friedrich Wilhelm Joseph von Schelling, *Ideen zu einer Philosophie der Natur*, Schellings Werke, ed. Manfred Schröter (Munich: Beck, 1956), pp. 82 and 170; *Von der Weltseele*, 2nd ed. (Hamburg: Friedrich Perthes, 1806), p. 67.

85. Goethe, *Wahlverwandtschaften*, p. 632.

86. Loc. cit.

87. Ibid., p. 268.

88. See notes to *Wahlverwandtschaften*, pp. 6, 27–53.

89. Ibid., p. 274.

90. Ibid., pp. 360 and 365.

91. Goethe, *Wahlverwandtschaften*, p. 366.

92. Ibid., p. 365.

93. Ibid., p. 367.

94. Loc. cit.

95. Goethe, *Wahlverwandtschaften*, p. 490.

96. In the notes to the *Wahlverwandtschaften*, p. 653.

97. Carl Friedrich von Weizsäcker, "Nachwort," *Die Einheit der Natur* , Hamburger Ausgabe (Hamburg: Christian Wegner, 1971), pp. 540–42.

98. Frederick Amrine and Francis J. Zucker, "Postscript: Goethe's Science: An Alternative *to* Modern Science or *within* It—or No Alternative at All?" in *Goethe and the Sciences: A Reappraisal*, ed. Amrine, Zucker, and Harvey Wheeler (Dordrecht: Reidel, 1987), pp. 373–88.

99. Weizsäcker, pp. 542–45.

100. Nicholas Boyle, *Goethe: The Poet and the Age*, vol. 2 (Oxford: Clarendon Press, 2000), pp. 92–93 and 202–6.

101. Goethe, *Wahlverwandtschaften*, p. 245.

102. Haus Jörg Sandkühler, *Friedrich Wilhelm Joseph Schelling* (Stuttgart: Metzler, 1970), pp. 63–67.

103. Schelling, "Ueber den wahren Begriff der Naturphilosophie" (1801), in *Ausgewählte Schriften*, ed. Manfred Frank (Frankfurt: Suhrkamp, 1985), vol. 2, pp. 1–35.

104. Joseph L. Esposito, *Schelling's Idealism and Philosophy of Nature* (Lewisburg: Bucknell University Press, 1977), pp. 31–46.

105. F. W. J. von Schelling, *Ideas for a Philosophy of Nature (Texts in German Philosophy)*, Errol E. Harris and Peter Hearth, translators, 2nd ed., Cambridge University Press, 1988; F. W. J. von Schelling, "On the World Soul," Iain Hamilton Grant, translator, *Collapse* 6:58–95 (2010).

106. Goethe, "Weltseele," Hamburger Ausgabe (Hamburg: Christian Wegner [1966]), vol. 1, pp. 248–49.

107. Schelling, *Ideen*, pp. 197–200.

108. Esposito, pp. 9–10.

109. Ibid., p. 10.

110. Ibid., pp. 269–70.

111. Ibid., p. 168. Schelling is not consistent on explanation. Sometimes he claims it for philosophy; at other times he surrenders it to science. See *Weltseele*, p. 11.

112. Esposito, pp. 68–77.

113. Loc. cit.

114. Newton, p. 943. See also Bernard Cohen's comments to *The Principia*, ibid., pp. 274–80.

115. Schelling, *Ideen*, p. 198.

116. Ibid., p. 62.

117. On chemistry, see Schelling, *Ideen*, p. 259; on electricity, ibid., pp. 144–46.

118. Ibid., p. 193.

119. Ibid., pp. 193–94. See also p. 259.

120. Ibid., p. 243.

121. Ibid., p. 243; *Weltseele*, pp. 11 and 46.—Strictly speaking dynamics is a branch of mechanics (as are kinematics and statics). Perhaps in deference to this, Schelling in *Ideen*, p. 259, calls mechanical dynamics "general dynamics" and the dynamics of contingency "special dynamics."

122. Schelling, *Ideen*, pp. 349–50.

123. Schelling, *Weltseele*, p. xii.

124. Ibid., p. 134.

125. Ibid., p. 135.

126. Loc. cit.

127. Ibid., p. 234. Kant, *Kritik der Urteilskraft*, Akademie-Ausgabe (Berlin: Georg Reimer, 1911), vol. 6, pp. 68–69.

128. Schelling, *Weltseele*, p. 235.

129. See his theories of light and of heat and heating, ibid., pp. 29–42 and 43–80.

130. Schelling, *Ideen*, pp. 121, 123, 139, 148, 153, 298, 304, 320; *Weltseele*, pp. 95 and 121.

131. Boyle, pp. 593–94.

132. See Schelling's additions to the second editions of *Ideen* and *Weltseele* (included in the editions referred to in note xcii).

133. Schelling, *Philosophie der Kunst, Ausgewählte Schriften*, vol. 2, pp. 185–86.

134. Schelling, "Fernere Darstellungen aus dem System der Philosophie," ibid., p. 167.

135. Robert J. Richards, "Nature is the Poetry of the Mind, or How Schelling Solved Goethe's Kantian Problems," in *The Kantian Legacy in Nineteenth Century Science*, ed. Michael Friedman and Alfred Nordman (Boston: MIT Press, 2006), chapter 3.

136. Richards, pp. 26–27.

137. Ibid., p. 3.

138. Ibid., pp. 10, 27, 9, 23, and 27.

139. Ibid., p. 27.

140. Goethe, "Wissenschaft und Erfahrung," p. 12 and his poem "Natur und Kunst," Hamburger Ausgabe (Hamburg: Christian Wegner [1966]), vol. 1, p. 245.

141. For more on high contingency, see my *Holding On to Reality* (Chicago: University of Chicago Press, 1999), pp. 105–13.

142. Kant, *Grundlegung der Metaphysik der Sitten, Akademie-Ausgabe* (Berlin: Georg Reimer, 1911), vol. 4, pp. 406–45.

143. Goethe, *Faust*, p. 17 (lines 289–92).

144. Schelling, "Ueber das Wesen deutscher Wissenschaft," *Ausgewählte Schriften*, ed. Manfred Frank (Frankfurt: Suhrkamp, 1985), vol. 4, pp. 11–28.

145. Ibid., p. 21.

146. Ibid.

147. Max Tegmark, "Parallel Universes," *Scientific American*, May 2003, pp. 40–51.

148. Frank J. Tipler, *The Physics of Immortality* (New York: Anchor Books, 1995).

149. Eric J. Chaisson, *Cosmic Evolution* (Cambridge, MA: Harvard University Press, 2001).

150. Brian Swimme and Thomas Berry, *The Universe Story* (New York: HarperSanFrancisco, 1994).

151. Robert Pack, *Before It Vanishes* (Boston: Godine, 1989).

152. Brian Greene, *The Elegant Universe* (New York: Norton, 1999), pp. 124–26 and 167–70.

153. Steven Weinberg, "A Model of Leptons," *Physical Review Letters* (volume 19, November 20, 1967), p. 1264 ff.). See also Xianhao Xin, "Glashow-Weinberg-Salam Model: An Example of Electroweak Symmetry Breaking," https://guava.physics.uiuc.edu/~nigel/courses/569/Essays_Fall2007/files/xianhao_xin.pdf

154. George Lakoff and Mark Johnson, *Metaphors We Live By* (Chicago: University of Chicago Press, 1980).

2

Ethics within Physics

FROM PHYSICS TO ETHICS TO PHYSICS

The great rival of ethics in explaining the structure of reality today is physics. Can we decree a truce of these rivals? Physics has nothing to say about ethics. Ethics has nothing to say about physics. Physics, of course, should be done ethically, and ethics is realized physically. But as theories, physics and ethics have no common areas of authority. Neither can tell the other how to proceed as a theory. They have nonoverlapping magisteria as Stephen Gould has told us.[1] Physics itself has two nonoverlapping magisteria, the magisterium of the small, quantum theory, and the magisterium of the large, relativity theory. They do overlap a little in relativistic quantum mechanics. But it's special relativity that has been reconciled with the quantum mechanics. General relativity and quantum mechanics have resisted reconciliation to the frustration and sometimes to the despair of physicists. In one way, however, the split between physics and ethics and between relativity theory and quantum mechanics is smoothed over and apparently healed.

There is a smooth path from the world at its largest to the world at its smallest. You can trace it by beginning at the middle, the midsize world of humans and their surroundings. The first to trace the path was Kees Boeke, a school reformer and teacher in Holland.[2] In the 1950s, he drew pictures for his pupils that showed a girl in a deck chair from above at ever greater distances, revealing the girl first in a courtyard, then in all of Holland, on the Earth, and so on until the universe at its largest scale was pictured. Returning to the girl, the view of the midsize world was shown at ever smaller scales—the hand, a blood vessel, molecules, down to the smallest particles then known.

It was natural to connect the pictures into the continuity of a film, and this has been done at least four times. All the films start from some restful or playful scene, a couple picnicking, children playing, a woman resting on a lawn, a boy in a boat with his dog. These settings are suffused with ethics. They are sketches of the good life—of rest, of care, of friendship, of pleasure, of celebration. The films depart from such a scene and return to it. It would be easy to reconfigure the films, having them begin at the small scale of the quantum world, ascend to the middle world of humans, and rise to the world at its largest where general relativity rules.

In such a journey we would not only move in apparently unbroken continuity from the quantum realm to the universe of general relativity, we would also move smoothly from physics to ethics and from ethics to physics, from atomic particles to the world where humans are at peace with themselves and one another which, to be sure, is also the world where humans struggle and suffer; and as the worldview rises above the earth, along with "its painful crudity and hopeless dreariness," the cosmic "contours apparently built for eternity" reveal themselves (to adopt Albert Einstein's words), the contours that are governed by the laws of general relativity—we're again in physics.[3]

Physicists have pointed out that reality is at one with itself, and so our best theory of reality should be too. "Quantum mechanics and general relativity are incompatible" Sean Carroll has said, "but nature is not incompatible with itself." And David Deutsch has similarly said that the fields of physics have to be connected "because the truth of the world has to be connected."[4]

COSMIC VIEWS

Does the physical cohesion of reality that emerges as we move from the smallest to the largest actually reveal a continuity between physics and ethics and so a moral cosmology? To contemplate what so unfolds is not an entirely recent enterprise. Blaise Pascal (1623–1662) invites his readers to consider a mite "with its minute body and parts incomparably more minute, legs with their joints, veins in those legs, blood in those veins, humors in that blood, drops in those humors, vapors in those drops" and to divide "those last things again." He then turns the readers' gaze to an "infinity of universes," and concludes: "Let him lose himself in wonders as amazing in their littleness as the others in their extension."[5]

For Pascal, a human as the "the middle of things," will be overcome by incomprehension and despair. And he adds: "the parts of the world are all so related and linked to one another that I believe it is impossible to know one without the other and without the whole." Similarly, a little later he says: "Since everything, then, is cause and effect, dependent and supporting, mediate and immediate, and all is held together by a natural though imperceptible chain which binds together things most distant and most different, I hold it equally impossible to know the parts without knowing the whole and to know the whole without knowing the parts in detail." Moral cosmology, however, does not require divine omniscience but rather intelligent acceptance of our condition. "This being well comprehended, I believe that one will hold oneself in repose, each in the state where nature has placed one." Insightful rest can be the focal center of a moral cosmology.

Contemporary cosmologists rightly take physics to provide the framework for moral cosmology if such a thing is possible. "By deepening our understanding of the true nature of physical reality," says Brian Greene, "we profoundly reconfigure our sense of ourselves and our experience of the universe."[6] But what this reconfiguration comes to, Greene does not say. Steven Weinberg's view has been melancholy: "The more the universe seems incomprehensible, the more it also seems pointless." On the next page, however, he may have left a seed for a moral cosmology: "The effort to understand the universe is one of the few things that lifts human life a little above the level of farce, and gives it some of the grace of tragedy."[7] In a later book, Weinberg returned to his remarks and to reactions of fellow physicists.[8] Most of them, it appeared, were puzzled by the very endeavor of searching for the universe to have a point. For his part, Weinberg professes nostalgia for the moral cosmology of the past. His remark, he said, was, "nostalgic for a world in which the heavens declared the glory of God." And a little further down the page he added: "It would be wonderful to find in the laws of nature a plan prepared by a concerned creator in which humans played a special role. I find sadness in doubting that we will."[9] Weinberg acknowledged that there are physicists who are as religious as they are competent. But he could not see a connection between their faith and their physics.

There are, however, physicists who return from a Kees Boeke tour cheerfully and hopefully. In *The Universe Story*, Brian Swimme (along with Thomas Berry, a theologian) adds the dimension of time to the spatial view of Boeke. Swimme and Berry invite us to celebrate the Big Bang as the "Primordial Flaring Forth" and its expansion into a life-enabling order that in time has produced life on Earth.[10] They see the universe story coming to a fork where we may take the destructive route of the Technozoic era or the responsible route of the Ecozoic era.[11] Eric Chaisson sees a pervasive and illuminating pattern of evolution in the universe, structuring the cosmic phases of radiation, matter, and life and so uniting the realms of physics, biology, and culture. The thread that runs through these regions is energy flow (divided by time and mass of the several phenomena). Chaisson's hope is that this "grand evolutionary synthesis would better enable us to understand who we are, whence we came, and how we fit into the overall scheme of things."[12]

Physicist Joel Primack and writer Nancy Ellen Abrams have come up with a venturesome and encompassing view of the cosmos (a view that's closest to the aspirations of this book) where humans become visible at the center of the universe. Our position is central with respect to time, size, and cognition. We are special also in being made of the rarest cosmic material. Seeing ourselves that way, Primack and Abrams suggest, we should be filled with joy and gratitude and a profound sense of responsibility for the welfare of the Earth.[13]

These several views are not incompatible with one another, but neither do they converge on a unified and principled moral cosmology. Are varied and optional proposals all we can hope for? Diversity without foundations is one problem of the available moral cosmologies. Another is the lack of a truly *cosmological* ethics. All of the implied or proffered moral conclusions (perhaps with the exception of Primack and Abrams's) that cosmologists have come up with are familiar. Students of environmental ethics have long urged stewardship of the earth and have done so without the need of cosmological guidance. A truly moral cosmology would imply moral insights that open up only within a cosmological perspective, and those insights should of course matter here and now.

THE FOUNDATIONAL ORDER

Perhaps such an ethics can be found if we start with the question: Is there a rigorous way of building an ethics from the ground up? And if so, what is *ground*? When you investigate the world ever more closely, then, no matter where you start, you eventually hit bottom as Kees Boeke did and as did the films inspired by Boeke. You inevitably end up at the ultimate building blocks and forces of reality, at the irreducible fermions and the force-carrying bosons. So, it's plausible to conclude, as Alex Rosenberg has concluded, that

> the basic things everything is made up of are fermions and bosons. That's it. . . . There is no third kind of subatomic particle. And everything is made up of these two kinds of things. . . . All the processes in the universe, from atomic to bodily to mental, are purely physical processes involving fermions and bosons interacting with one another. [14]

It doesn't matter where you start, whether at a salt crystal or a bit of organic matter, you end up with fermions and bosons in a world that is lawfully crisp and clear. Everything is reducible to the world of quantum physics, including the things and forces of the midsize world—buildings and human beings, the force of the wind and of a campaign. It seems reasonable to conclude that the confusing and messy midsize world is thus reducible to the clear and clean world of quantum physics.

The way to make the fundamental order of the ultimate building blocks and forces prevail in the middle world of humans is to reconstruct that world from the bottom up. That idea is supported by a foundational order. When you follow Kees Boeke's ascent, the objects you see exhibited in sequence, up to the human level, correspond to a sequence of disciplines (starting at the bottom of table 2.1):

Table 2.1. Sequence of disciplines, based on Kees Boeke's *Cosmic View: The Universe in 40 Jumps.*

OBJECTS	DISCIPLINES
Societies	Sociology
Organisms	Biology
Molecules	Chemistry
Atoms	Physics
Structures	Mathematics

Ascending in the order of disciplines, we can say that one has transformed the next. Mathematics began to transform physics when Galileo said of the book of the universe:

> It is written in the language of mathematics, and its characters are triangles, circles, and other geometrical figures, without which it is humanly impossible to understand a word of it; without these one is wandering about in a dark labyrinth.[15]

As for chemistry, it was the employment of the language of atoms by John Dalton in the early 1800s that proved Kant wrong. Detailed understanding of atoms, their particles, and their chemical properties had to await the quantum theory of the first half of the twentieth century. The landmark event of the transformation of biology by chemistry was the discovery of the structure of DNA by James Watson and Francis Crick in 1953. And the first influential proposal of how biology should transform sociology, once we assume that the social is at the bottom biological, was Edward O. Wilson's 1957 *Sociobiology*, followed by the upsurge of evolutionary psychology.[16]

These disciplinary transformations suggest that the objects, corresponding to the disciplines, are founded one on the other, atoms on structures, molecules on atoms, and so on up. This foundational order in turn suggests that the luminous order of mathematics gets complicated and perhaps concealed as we ascend the foundational order, but that with rigor and care it can also be made visible and so enlighten the world entire, including the moral conduct of humans. There is a long tradition in philosophy, beginning with Plotinus (205–270) that seeks to demonstrate the unfolding of a great order from simplicity to complexity, often called "the great chain of being."[17] A relatively recent and widely admired attempt to make good on the program with formal rigor is Rudolf Carnap's *Der logische Aufbau der Welt* of 1928 (translated as *The Logical Structure of the World*).[18]

Trust in the rigorous and illuminating force of the foundational order may be strengthened by the inescapable reducibility of all there is. Start with any object in the universe or start with the universe itself, examine it ever more closely, and you end up at the irreducible quantum level. Reducibility is sometimes seen as a threat to the familiar world we cherish. Reduction, it is feared, shows that the beloved midlevel world is not real, that it is the fragile product of our imagination and conceptions. The reply to the threats are "demonstrations" that a person or a sunset is not reducible via the descent to the foundational order. But by and large these apprehensions are unfounded and the demonstrations unsuccessful.

REDUCTION AND EMERGENCE

The shortcomings of reductions are not that they have to stop and need to be abandoned at some high level of the foundational order, but rather that reductions progressively lose what is thought they reduce. Consider the steps of reduction in Philip and Phylis Morrison's *Powers of Ten*.[19] At a distance of one meter you see a young man napping after a picnic. At a tenth of a meter, you have lost the person and see only his hand. At a tenth of a millionth of a meter you see the coils of the man's DNA. It's his DNA and the last trace of his identity because at the next step, at a hundredth of a millionth of a meter, all you see are the building blocks of the DNA, the nucleotides. At this level what shows up is a carbon atom that could have come from inorganic matter or from coal.

Thus, the reduction more and more loses what it was supposed to reduce. Would it be possible to reduce an entire human being to atoms so to comprehend a person at close to the most fundamental level of reality? Assume we analyze a person step by step and try to record the steps of our analysis on a map—first a picture of the body, then of its organs, next of the cells of the organs, of the molecules of the cells, and finally of the 7×10^{27} atoms that compose the average human; how much terrestrial surface would the final record of our analysis take? Allow for one square inch per atom. Divide the number of atoms in a human by the number of square inches on Earth (7.93×10^{17}):

$$7 \times 10^{27} \div 7.93 \times 10^{17} = 1.261,034,047,920 \text{ trillion}$$

More than a trillion earths would be needed to accommodate the atoms of the average person on the assumption that a square inch is needed to register one atom. The number is astronomical in size and cosmic in significance. It tells us that the reduction of the average person to atoms is incomprehensible. It would not make much of a difference if we required only a square centimeter per atom. A square inch contains 6.4516 square centimeters. Hence the number of earths needed on that score would still be 1.5907642×10^{13}. It's only when we go down to the miniscule scale of square nanometers that less than two thousandths (0.00159066553 to be exact) of the terrestrial surface is needed for the atomic map. That's 270,254 square miles, roughly the size of Morocco.

Atoms, of course do not constitute the most basic level of the foundational order. They are composed of the fermions and governed by the bosons of the quantum level at a scale of one Fermi. Impossible as it is to keep track of all ultimate particles of a human body, the fact that there are ultimate objects (the fermions) and ultimate laws of construction (represented by the bosons) makes you think that the rigorous order of the toy universe obtains in the actual world as well, albeit in a more elaborate version. And although the sheer number prevents an unfolding of the actual world as perspicuous as the unfolding of the toy universe, ours may still be a constructive world in an important sense: At bottom, it is rigorously constructed from objects and laws. Perhaps the philosophers of the great chain of being, Plotinus (205–270) and Leibniz (1646–1716) among them, were right that if only we looked carefully enough, we would see the all-pervasive order of the world; we would see the same order in physics and in ethics.[20] This conjecture would be substantiated if we could explicitly ascend the foundational order, showing in perfect detail how atoms are composed of fermions and bosons, molecules of atoms, organisms of molecules, and societies of organisms.

The how of this composing would be spelled out in laws of increasing specificity, each being a special case of its underlying law. This ascent, however, runs into several insuperable obstacles.[21] First and surprisingly, explicitness and rigorous detail get shrouded at the quantum level already. There are processes where we know the beginning and the end, but the in-between is hidden from us. So also with chemical reactions; and needless to say, biological and social processes are lawful in only a rough and ready way. Second, there are rigorous higher-level laws that demonstrably are not reducible to more fundamental laws.[22] Third, there are events where a general law can become more specific in more than one way. Why in fact it becomes specific this way rather than that is unknowable. There are, finally, similarly inscrutable conditions where the course of development rests on a fine ridge; and why it tumbles down one slope rather than the other with an outcome hugely different from what it had been otherwise, is simply beyond knowing.[23]

It's often hard to say whether the capriciousness of the world is woven into the fabric of reality or whether it is just a reflection of our ignorance, curable in principle by more research. That does not mean that in the recesses of caprice there lurk mysterious objects and forces (physical lawfulness is closed). It's all fermions and bosons. Nor does it mean that it's impossible or pointless to focus minutely on parts of some object. It's both interesting and useful to see how neurons connect at the point of contact (a synapse) and how a chemical substance (a neurotransmitter) flows from one brain cell to the other. But what does it mean?

What emerges from fermions and bosons is our articulate world of colors and shapes, of things and person, of situations and events. It might have been otherwise. Fermions and bosons could have issued in chaos or monotony. The contingent emergence of a world of commanding presence is, along with lawfulness, the great fact of reality. But how and when did emergence take place? As philosophers think of it, emergence is the conceptual process of ascending the foundational order in our understanding. Emergence is the connection, or stands for the search of a connection, between what Winfred Sellars has called the scientific image of physics and the manifest image of the everyday world.[24] Historically and actually, emergence is the evolution of our many-sided world from the uniformity and density of the Big Bang.

So, our familiar world and its moral complexion have conceptually and cosmologically emerged from the processes that are described by quantum and relativity physics? So, it has. But just as we have failed to see in emergence how quantum mechanics and general relativity hang together, we have yet to see whether in emergence physics informs ethics uniquely and significantly.

NOTES

1. Steven Jay Gould, *Rocks of Ages: Science and Religion in the Fullness of Life* (New York: Ballantine Books, 2011).

2. Kees Boeke, *Cosmic View: The Universe in 40 Jumps* (New York: John Day, 1957).

3. Albert Einstein, "Principles of Research," https://www.site.uottawa.ca/~yymao/impact/einstein.html, accessed June 4, 2021.

4. Sean Carroll, "Why Does the Universe Look the Way It Does?" *The Universe*, ed. John Brockman (New York: Harper, 2014), p. 102; David Deutsch, "Constructor Theory," loc. cit., p. 365.

5. Blaise Pascal, *Pensées*, ed. Victor Giraud (Paris: Rombaldi, no date [1935], pp. 35–42 (Number 72). My translations.

6. Brian Greene, *The Fabric of the Cosmos* (New York: Alfred A. Knopf, 2004), p. 5.

7. Weinberg, *The First Three Minutes* (New York: Basic Books, 2nd ed. 1993 [1977]), pp. 154 and 155.

8. Weinberg, *Dreams*, pp. 255–57.

9. Loc. cit., p. 256.

10. Brian Swimme and Thomas Berry, *The Universe Story* (New York: Harper Collins, 1992), title of ch. 1, pp. 17–29.

11. Ibid., pp. 249–50.

12. Eric J. Chaisson, *Cosmic Evolution* (Cambridge, MA: Harvard University Press, 2001), p. x.

13. Joel R. Primack and Nancy Ellen Abrams, *The View from the Center of the Universe* (New York: Riverhead Books, 2006).

14. The dots are Rosenberg's. Alex Rosenberg, *The Atheist's Guide to Reality* (New York: Norton, 2011), p. 21.

15. Galileo, *Assayer*, p. 184.

16. E. O. Wilson, *Sociobiology: The New Synthesis* (Cambridge, MA: Harvard University Press, 2000 [1957]).

17. Arthur O. Lovejoy, *The Great Chain of Being* (Cambridge, MA: Harvard University Press, 1964 [1936]).

18. Rudolf Carnap, *Der logische Aufbau der Welt* (Hamburg: Felix Meiner, 1998 [1928]).

19. Philip and Phylis Morrison, *Powers of Ten: About the Relative Size of Things in the Universe* (New York: Scientific American Books, 1982).

20. See Arthur O. Lovejoy, *The Great Chain of Being* (Cambridge, MA: Harvard University Press, 1998 [1936]).

21. Gordon G. Brittan Jr., "Explanation and Reduction," *The Journal of Philosophy*, vol. 67 (July 1970), pp. 446–57. For Carnap's failure on his own terms, see Nelson Goodman, *The Structure of Appearance*, 2nd ed. (Indianapolis: Bobbs-Merrill, 1966 [1951]), pp. 151–87.

22. P. W. Anderson, "More is Different: Broken Symmetry and the Nature of the Hierarchical Structure of Science," *Science*, August 4, 1972, vol. 177, no. 4047, pp. 393–96.

23. Stephen H. Kellert, *In the Wake of Chaos: Unpredictable Order in Dynamical Systems* (Chicago: University of Chicago Press: 1993).

24. Wilfrid Sellars, "Philosophy and the Scientific Image of Man," *Frontiers of Science and Philosophy*, ed. Robert Colodny (Pittsburgh, PA: University of Pittsburgh Press, 1962), pp. 35–78.

3

Background Conditions

THE ENLIGHTENMENT

If there is a moral cosmology, we have to learn to see it and to sense it. The possibility of recognizing a moral cosmology opens up only if we are able to move beyond the modern conception of how we today appropriate our world. It's a conception that has split into two parts, one cognitive, the other ethical, or as Kant has it, one theoretical, the other practical. That dual conception was born suddenly, a matter of a few centuries. In retrospect we can discern a scatter of factors that converged and combined to produce it. But no one could have predicted the development. The emergence of modern culture is one of the great contingencies of history.

It's always possible, of course, to find a point of view from which the sequence of episodes falls into a pattern like the variations of a theme or the values of a variable, the theme or variable being competition among humans or the search for comfort or security or power. It's not incorrect, of course, to see and to detail such patterns, but so to analyze history is to be governed by one of the conceits of modern history rather than to discover its moral cosmology. Premodern cultures, as a rule, thought of themselves as exclusively unique, as centered on Earth and, when relative population density increased past kinship, surrounded by inferiors or by enemies rather than by alternatives. To take the view from nowhere of all the cultures there are and have been is to take a new historical and cultural position, a position that was born of modernity.

One way of grasping the basic transformative force of modernity is by way of the term that has been given to its founding cultural event—the Enlightenment. It's usually seen as an intellectual process, the articulation of the critiques and visions that ushered in the modern era. I don't want to suggest, however, that ideas were the drivers of the tangible culture. Even in the seemingly plausible cases where a speaker's ideas did move people to action, the conditions on the ground had to be favorable. More often, ideas have been clear and helpful anticipations or reflections, rather than actual causes, of epochal events. The line between insights and causes is fuzzy, of course. If you know what you're doing, you're more likely to be effective.

At any rate, I will use the term and the notion of the Enlightenment to illuminate the basic and pervasive event that gave rise to our prevailing notions of knowledge and ethics. The concept of the Enlightenment captures both the fracturing and the revealing forces of modernity. To enlighten someone is to break down that person's prejudices and superstitions. "*Enlightenment happens,*" Kant said, "*when humans leave the tutelage that they had willingly submitted to.*"[1] But to enlighten is also to reveal previously unseen forces and structures.

As far as cosmology is concerned, the Enlightenment silenced or swept away the moral eloquence of the universe. "Tis all in pieces, all coherence gone," said John Donne in 1621.[2] In 1788, Friedrich Schiller said:

> Where now, as our sages say,
> A soulless ball of fire turns,
> His golden chariot once
> Helios steered in quiet majesty.[3]

"All that is grounded and at rest is vaporized, all that is holy is profaned," said Karl Marx and Friedrich Engels in 1848.[4] In 1917, Max Weber noted the common belief that there were no longer any mysterious incalculable powers, that everything could in principle be conquered through calculation; and he concluded: "That, however, means: The disenchantment of the world."[5]

In sweeping away one thing, the Enlightenment revealed another—a space that is unimaginably huge, has no center, and makes the earth appear miniscule and adrift. There is no definitive here anymore. Nor is there a decisive now. The present moment you share with something or someone else depends on the direction you take and on the speed you travel relative to that thing or that person. Finally, the cosmos may have had a beginning, but it has no end. Its structure diffuses endlessly into emptiness.

ORIENTATION IN AN ENLIGHTENED WORLD—KANT

Culturally, relational space and its regions rest on landmarks large and small and the relations among them. The village of Törbel in the Valais canton of Switzerland may have been one region of a Swiss mercenary and the battlefield near Novara in Northern Italy a landmark. (The battle of Novara took place in 1513 and the Swiss mercenaries prevailed.) Here was home, there was danger and victory. As these landmarks and their regions began to lose their orienting force, so did the space of relations they had anchored. How then was orientation possible? Here is Kant's answer:

> In bodily space, because of its three dimensions, it is possible to imagine three planes which intersect one another entirely at right angles. Since we know everything through

our senses only insofar as it is related to us, it is no wonder that we take the first basis to generate the concept of regions in space from the relation of these intersecting planes to our body.[6]

Three-dimensional coordinate systems are basic to modern physics. Look at the way a contemporary introduction to classical physics presents it:

> To describe points quantitatively, we need to have a coordinate system. Constructing a coordinate system begins with choosing a point of space to be the *origin*. Sometimes the origin is chosen to make the equations especially simple. For example, the theory of the solar system would look more complicated if we put the origin anywhere but at the Sun. Strictly speaking, the location of the origin is arbitrary—put it anywhere—but once it is chosen, stick with the choice.
> The next step is to choose three perpendicular axes. Again, their location is somewhat arbitrary as long as they are perpendicular. The axes are usually called x, y, and z, but we can also call them x1, x2, and x3.[7]

Evidently you can define a coordinate through three planes or through three axes, x, y, and z. Conventionally we arrange them so that x is the axis of width, y of height and z of depth. Kant describes the planes informally. What he calls "the horizontal plane" is the one that contains x- and y-axes. The one that "yields the basics of the difference between the right and the left side contains the x- and z-axes. Finally, the plane that gives us "the concept of the front and the back side" contains the y- and z-axes.

A coordinate system provides for orientation, it allows you to "describe points quantitatively." Something has an unambiguous position in such a system and unambiguous relations to everything else. But where is the system itself located? Kant suggests that the center or origin of the system is located in the self or the ego. Crucially, it is the distinction between my left and my right that enables me to distinguish regions and objects in space. Important for our purposes, the determination even of cosmic regions and directions requires the left-right distinction, says Kant. (One could argue that the ego merely lends orientation to a coordinate-system whose location in absolute space is given to the ego. I will proceed on the assumption that the ego does both—determine the location of and lend orientation to the coordinate system.)

The core of Kant's argument for absolute space is a discussion of certain pairs of things that resist identification of one thing as distinguished from the other by means of spatial relations only. Kant calls them "incongruent counterparts." In mathematics they are called enantiomorphs. They are composed of parts, and the parts have relations to each other. Here's an illustration why the left-right distinction, called handedness or chirality, that the self brings to the world, is needed to identify an enantiomorph and to distinguish it from its counterpart. Say you have injured your hand, and you phone your mother for sympathy. To explain where on your hand you suffered your injury, you tell your mom where the wound is located in relation to your thumb and your middle finger. You can describe things in whatever detail, but your mother won't know which hand it is that got injured until you employ the left-right distinction or one that is equivalent to it. Kant's essay of 1768 suggests that what allows for orientation in space is a coordinate system that is centered on the self.

Its origin is the I: *Ego est origo*. And being at the center of space, the self provides the crucially orienting distinction between right and left. We can assume that for Kant chirality stands in for the equally important distinctions between up and down and front and back.

Eighteen years later, in 1786, Kant returned to the issue in an essay titled, "What Does it Mean to Orient Oneself in Thought?"[8] He picks things up where he has left them—the problem of orientation in space and the solution that handedness furnishes. Having meanwhile published his masterpiece, *Critique of Pure Reason* (1781), Kant is much more pointed in claiming that chirality is brought to reality by reason and is not taken from reality by experience.[9] He further expands the orienting force of reason to thought in general and, in fact, to ethics. Reason is the origin of the necessary conditions of knowledge and of the moral law. It's a bold as well as tenuous connection between what we can know and how we should act—reason in its purity is the sole source of universality and necessity. As we have seen, the tie snapped when Kant proclaimed the separation of cosmology from ethics. In a footnote Kant ties this essay to his earlier one, "What is Enlightenment?" (1784): "*To think for oneself* is always to look for the supreme acid test of truth in oneself (i.e., in one's own reason); and the maxim always to think for oneself is *Enlightenment*."[10]

KINDS OF KNOWLEDGE

What concerns us here is what Kant captured in the modern cognitive attitude, in the common view of how we should come to know, viz., autonomously, free of assumptions and prejudices. Hans-Georg Gadamer calls this "The discrediting of prejudice by the enlightenment."[11] This attitude comes to the fore in admonitions to set aside or even shatter established frameworks, to be innovative and even disruptive in our approach to the world, to give things a fresh look, to accept only what is plainly given. Information technology has given this attitude particular panache—it's stupid to burden one's mind with information when fallible memory has been replaced with the instantaneous and autonomous retrieval of precise information from the Internet. Knowledge, however, is crucial to world appropriation, to being in the world fully and well.

There is an abundance of scholarly material about knowledge, and there are three ways of dealing with it. The first and worst is to gather it all or try to do so. The result is a heap of observations, concepts, and theories. You have all there is to know about knowledge before you, but it just sits there and defies understanding. The second is to construct a continuum of concepts or theories extending between two prominent and extreme positions; or to construct a matrix if you're more ambitious; or a three-dimensional grid if you're most ambitious. In any of these structures, every piece in the heap of materials finds its place and is related to every other piece, much as are the objects of the constructive world. But those structures, however ingenious and orderly, are still lifeless and uninspiring. So, the third way, the one we should strive for, is animated by a moral concern that allows the relevant materials to coalesce into a building that makes sense and invites inhabitation. Surely a moral cosmology would have to be such an edifice. But to build it we have to do some sorting of materials before we can hope to join them helpfully.

There are scholarly disciplines that deal with knowledge. There is epistemology in philosophy. A more recent discipline is cognitive science which combines many more disciplines—logic, philosophy of mind, psychology, linguistics, information theory, neurophysiology, etc. I will draw on them liberally and selectively. What the multiplicity of disciplines tells us is that knowledge is a rich and complex phenomenon, always exceeding the scholarly reach and grasp. So is the brain where knowledge is housed. The brain at once fascinates and frustrates researchers. It fascinates because it is so crucial and so highly and finely organized. It frustrates because it always acts as a whole more or less, all eighty-six billion nerve cells and their trillions of connections more or less. There's no hope of mapping it all as we have seen, and of tracing just the crucial process of what is happening in your brain when, say, you are reading this.[12] The mind, moreover, is inseparable from the body, and the body from the world.[13]

In 1949 Gilbert Ryle introduced the distinction between knowing how and knowing that[14] (see Table 3.1). It has been rephrased and retraced in many ways. As the last pair shows, memory overlaps with the settled and enduring knowledge that's called long-term knowledge, the major concern of what follows.

Table 3.1. Distinctions between "knowing how" and "knowing that." Based on Gilbert Ryle's *The Concept of Mind*.

Knowing how		Knowing that
Skill	→	Content
Procedure	→	Propositions
Habit memory	→	Semantic memory

My favored versions of the distinction are the ones between procedural knowledge and substantive knowledge and between knowing how and knowing that. Together with my criticism of unprejudiced knowledge they will be my point of departure. My goal is a notion of background knowledge that I will develop bit by bit, so here we go.

BACKGROUND KNOWLEDGE

There is a perfectly valid and common sense of "prejudice," such that prejudices should be avoided. They should be when "prejudice" means erroneous, misleading, or mischievous background information. Such prejudices lead to errors of judgment and, worse, to unfair or immoral claims and actions. But it's one thing to fight against prejudice in this sense and another to reject all knowledge that's prior to perception or judgment. Hans-Georg Gadamer (1900–2002) uses "prejudice" (German *Vorurteil*) as a term for the prior knowledge that's indispensable for understanding.[15] I will argue that Gadamer is right in substance, but I will use background knowledge where he uses prejudice.

Here's an example to illustrate the force of background knowledge. You're sitting at the table in the seminar room, and you're facing the cream-colored plastered wall that separates the seminar room from the hall. You can't see the hall, of course, so a truly justified statement would be: I see a cream-colored surface. Cream-colored? The color that's somewhere between white and yellow? But what justifies your bringing this color scheme to your perception rather than another or any at all? And surface? Something flat that's between a line and a cube? But what entitles you to bring Euclidean geometry to your perceptual judgment rather than Lobachevskian or any geometry at all? Once you have purged your perception of all background knowledge, all you're justified in claiming is: "I see something." But that statement tells me nothing about what you are seeing. What you truly see when you look up from the table in the seminar room is a plaster wall in a brick building on a campus in Missoula, in Montana, in the United States, on the North American Continent, on planet Earth, in the solar system, in our galaxy, amid many galaxies, among clusters of galaxies, in our universe, and perhaps among countless other universes. Obviously, I have just scaled the Boeke ladder from the midsize world on up as far as the ascent goes. Once more we've come to the leading question of this essay: Do ascending and descending help us to appropriate the world in an intrinsically moral way?

Just to hint at a possible answer, I'd like to suggest that the background of knowing-that morally informs our knowing-how and our being in the world. Knowing that this is a wall tells you that you are sheltered from the comings and goings of the hall and able to concentrate on the discussion about to begin. Seeing that the door is unlocked tells you that Leslie, who is always late, will slip in without having to knock. Knowing that the seminar is taking place under the auspices of the University lets you participate in the seminar proceedings with a seriousness that is underwritten by your confidence that what you say will be recorded in a way that clears a path for your future.

Thus, the background of your situation, present through your background knowledge informs your being in the world and the quality of your life. The background has moral force. It makes for a good life, and of course it could fail to do so. That the good life is at issue here seems clear to me. It's far less clear whether the norms of contemporary standard ethics help us to measure the moral quality of life within its background.[16] And whether knowledge of the deeper and wider cosmic contexts of life deepens and expands the goodness of life—whether there can be a moral cosmology—remains to be seen.

As far as the background of our know-how or procedural knowledge is concerned, we're the heirs of a long and rich evolution of processes that are almost entirely automatic and governed by the autonomic nervous system—heart rate, respiratory rate, and digestion are examples. We can influence them through meditation, exercise, and diet, but in the main these processes go on by themselves. Then there are skills that have a strong evolutionary and therefore genetic foundation, and on that basis, we acquire such skills as walking and talking. We do so with little instruction but much in need of the right social circumstances.

Finally, we are born with a basic understanding of how the world hangs together and how it works. This understanding is sometimes divided into folk versions of scholarly disciplines. Our folk physics tells us that solid objects cannot pass through each other and will not easily come apart. They are subject to gravity and require force to be moved. Folk biology makes us distinguish between living and nonliving things and makes us understand that living things are divided into natural kinds and species. Folk psychology enables us to get the meaning of smiles and to anticipate a person's reactions to kindness or hostility. This native and naive knowledge is rough and ready, subject to exceptions, and yet indispensable to our being in the world. Its procedural and substantive parts go hand in hand. Knowing what a fist-sized rock is like, we know how to move it with our foot. Knowing the difference between cats and dogs, we will respect cats and love dogs. There is more to be said about background knowledge—its presence in daily life, the way it pervades and colors everything.

NOTES

1. Kant, *Was ist Aufklärung?* (Hamburg: Felix Meiner, 1999 [1784]), p. 20 (my translation). Kant's emphasis (here in italics).

2. John Donne, *The Anniversaries*, Frank Manley, editor (Baltimore: Johns Hopkins University Press, 1963) pp. 73–74.

3. Friedrich Schiller, *Werke*, ed. Arthur Kutscher (Berlin: Deutsches Verlagshaus Bong & Co., no date [1788]), volume 1, p. 57. My translation.

4. Karl Marx and Friedrich Engels, *Manifest der kommunistischen Partei* (Hamburg: Argument Verlag, 1999 [1848]), pp. 47–48. My translation.

5. Max Weber, *Wissenschaft als Beruf* (Stuttgart: Reclam, 1995 [1917], p. 19. My translation.

6. Kant, "Von dem ersten Grunde," pp. 994–95.

7. Leonard Susskind and George Hrabovsky, *The Theoretical Minimum* (New York: Basic Books, 2013), p. 15.

8. Kant, "Was heißt: sich im Denken orientieren?" in *Was ist Aufklärung? Ausgewählte kleine Schriften*, ed. Horst D. Brandt (Hamburg: Felix Meiner, 1999), pp. 45–61.

9. Kant, *Kritik der reinen Vernunft*, ed. Raymund Schmidt (Hamburg: Felix Meiner, 1956 [1926]). First *as ist Aufklärung? Ausgewählte kleine Schriften*, ed. Horst D. Brandt (Hamburg: Felix Meiner, 1999 [1786]), pp. 45–61.

10. Kant, "Was heißt: sich im Denken orientieren?" p. 60, footnote 1 in "Beantwortung der Frage: Was ist Aufklärung?" *Ausgewählte kleine Schriften*, pp. 45–61.

11. Hans-Georg Gadamer, *Truth and Method*, translation edited by Garrett Barden and John Cumming (New York: The Crossroad Publishing Company, 1985 [1960]), pp. 241–45.

12. Stefan Theil, "Trouble in Mind," *Scientific American,* October 2015, pp. 36–42. Ed Young, "Will we ever . . . simulate the human brain?" BBC, February 8, 2013. The dots are the author's.

13. See my "Mind, Body, and World," *Philosophical Forum,* volume 8 (1976), pp. 68–86.

14. Gilbert Ryle, *The Concept of Mind* (New York: Barnes & Noble, 1968 [1949]), pp. 25–61.

15. Hans-Georg Gadamer, *Wahrheit und Methode* (Tübingen: Mohr Siebeck, 2010 [1960]), pp. 270–95.

16. See my *Real American Ethics* (Chicago: University of Chicago Press, 2006).

4

Ordinary Cosmology and Mathematics

FROM FOLK COSMOLOGY TO ORDINARY COSMOLOGY

There is also a folk cosmology. It is centered on the here and now, the place where we are located and the time we live in. Spatially, there is the vault of the sky above us and the foundation of the earth below us, with what's close being familiar and safe and what's far being strange and perilous. Time is structured by the cycles of day and night and of the seasons. Folk cosmology has always been a moral cosmology. The sky is the realm of the ruling and luminous forces humans have to submit to. Earth is supporting and nourishing, but it also harbors hidden and treacherous creatures. Within the common cosmology, there's a personal version, realized and centered in each of us. It's evident in the metaphors we live by.[1] Up is good, down is bad. Front is public, back is private. Right is handy, left (with apologies to lefties) is awkward. This micro cosmology is reflected in the single-family house where the front yard and facade are formal and the back yard informal and often messy. The public rooms—living room and dining room—are toward the front; the private rooms—bedrooms and bathrooms—are in the back of the house; the kitchen sits in the middle.[2]

We are heirs to the rudiments of the traditional version of folk cosmology, and it is part of the background knowledge that enables us to appropriate and negotiate our world. But since the Enlightenment, traditional cosmology has more and more been transformed by science and technology. Kant's two essays on orientation are prescient and telling. Orientation is needed and centered in the person and a person's up and down, left and right, front and back. But it's an orientation that has been purged of its animating life of good and bad, skillful and inept, public and private. It is formal with minimal residues of fully being in the world. The result of the fading of tradition and the rise of science and technology I will call ordinary cosmology. It's the cosmology of the ordinary person in an advanced technological society such as ours. I will use traditional cosmology as a backdrop to bring ordinary cosmology into relief, and eventually I will use ordinary cosmology as a backdrop for a normative cosmology, one that is knowledgeable and has cosmological moral force. At any rate, that's the aspiration. To articulate ordinary cosmology, we need an articulate conception of the ordinary person. Is such a thing available? Who or what is the woman or man in the street? What do ordinary people know? How do they see the world? You have to remember that each of us lives in the ordinary world and knows what ordinarily happens. We are familiar with the furniture of ordinary life and know what to expect from our fellow citizens when they use and navigate it. We can anticipate broadly what people will say and do in ordinary circumstances.

It is important to stress that it is the ordinary background knowledge of ordinary cosmology that concerns me. It's equally important to note that the concern with background knowledge is not a matter of applying a crisp theory to a messy subject. Rather as we delve into the particulars of ordinary people's ordinary cosmology, our general conception of background knowledge will gain in articulation and force.

Ordinary cosmology shares with contemporary culture the loss of absolute frameworks and orienting centers. That loss is shared in turn with the natural and social sciences. Cosmological physics is in a state of crisis. Chemistry has more and more disassembled and reassembled the materials we find in nature. We have gotten used to ever new materials with novel properties. The boundaries between chemistry and biology have been breached by biochemistry and microbiology. The dark fortress of cancer has not been conquered by these scientific advances, but some of the outer fortifications have been taken. The spirit of dissolution and reconstruction has been cheered, wrongly I'm convinced, in its attacks on aging and mortality, and its predictions that Artificial Intelligence will equal and in time surpass ours have been applauded too. In sociology, the shattering of the prejudices of gender and ethnicity has been part of the glories of the Enlightenment. But the stabilities of monogamy, of the family, and of democracy have come under attack also, with injurious consequences, I am sure.

For the most part, however, cultural pointlessness has not led to a chaos of individual autonomies. It's been contained, not by the common moral law, as Kant has proposed, but by a substructure that provides in concrete and endlessly varied forms what GPS furnishes broadly for space and time. We call the most basic part of that substructure the infrastructure. Layered on it are the structures of production, transportation, and commerce, some of them financed and all of them regulated by government.

So now we have before us the pieces that compose ordinary cosmology. Its crucial region is defined by technology. Spatially it is delimited by the shell that is traced by satellites of information and communication technology (ICT). On the ground it is shaped by the layers of infrastructure that together afford the structure of daily orientation. Daily life is tightly constrained by this substructure. We obey traffic signs and lights. We show up at the airport at the appointed times. Importantly, we show up at work to maintain and advance the technological structures through our labor.

Within the shell of technology, ordinary life shows the bequests of tradition—the times when we eat, the mores of socializing and of marriage in particular, the lay-out of our homes and cities, the places where we learn and where we worship. The shell itself shows vestiges of tradition, of traditional cosmology—the heavenly up and the earthly down, the light of day and the burdens of darkness at night, the center of family and familiarity and the periphery of the foreign and suspicious. Still, ordinary life and its boundaries are being eroded by the winds of liberation and the floods of pointlessness. These processes threaten the ordinary sense of orientation.

Ordinary cosmology shows affinities with science. With quantum mechanics it shares the ambiguities of reality and their resolutions, the many worlds that branch out from the divisions of the world that is and the worlds that might have been. It shares the instantaneity of entanglement interactions, no matter the distances. With (special) relativity theory, ordinary cosmology has in common the disappearance of absolute frameworks and the remaining stability of events and their connections. These are kinships that general education should have made known as parts of the foundational order. But for the most part, even when one or more layers of that order are well-known by scientists, if not by the general public, the order has little moral force. Ordinary cosmology is a fabric of tradition and technology. It's a small tapestry, however, and it's beginning to unravel.

MATHEMATICS AND ORDINARY COSMOLOGY

To give my account some cohesion I will follow the foundational order in drawing the lineaments of ordinary cosmology and begin with mathematics. It has two different aspects. One is of mathematics in its purity. As such it reveals abstract, rigorous, and intricate structures and those familiar with them, the mathematicians, often call them beautiful. Those mathematicians will tell you that doing pure mathematics is very hard work. Most also tell you that they discover those structures rather than invent them, and that the conclusive discovery or revelation is a moment of overwhelming joy. Andrew Wiles was moved to tears when he recalled the "revelation" of the crucial piece that was needed to prove Fermat's Last Theorem.[3]

"Discovery" and "Revelation" suggest that mathematical structures exist in their own right in a realm beyond the tangible universe. How we should think of that realm is a controversial question. But if there were no such realm, then no discoveries could be made and no revelations could come to pass. The other aspect of mathematics is its applicability to the observable and tangible world. Applicability would not be surprising if physics and mathematics progressed in locked steps as they did when Newton came up simultaneously with calculus and his laws of motion.

But in the case of general relativity and quantum mechanics, Einstein, Heisenberg, and Schrödinger applied largely available mathematical structures to capture physical lawfulness, structures that had been previously discovered as interesting and beautiful in their own right—a surprising fact that led Eugene Wigner to speak of "The Unreasonable Effectiveness of Mathematics in the Natural Sciences."[4] Numeracy is to mathematics as literacy is to reading and writing. The numeracy of pure mathematics is hardly taught at all in the United States. The emphasis is predominantly on application. Pedagogically, it is wise to use applications and illustrations widely. But if education in mathematics stops short of at least introducing students to the preternatural realm of pure mathematics, it deprives students of a beautiful insight. While pure numeracy is lacking, applied numeracy is deplorably low in this country, nearly the lowest among the developed nations. Ordinary people have trouble computing averages and percentages, estimating probabilities, understanding statistics, and spotting errors and deceptions in quantitative reasoning.

And yet we live in a world that's informed by mathematics, and so we should expect that our ordinary sense of being in the world has been so informed as well. Where can we hope to find evidence that this is so? We can take our clues from Thorstein Veblen and Max Weber. Veblen's is the more general hint. We should follow him in realizing that our insights have to be *"drawn from everyday life, by direct observation or through common notoriety."*[5] Weber's observation is more specific. In 1917 he already put his finger on the crucial point:

> The increasing intellectualization and rationalization, then, do not indicate an increasing general knowledge of the conditions of life one is subject to. They rather indicate something else—the knowledge or belief that one could, if only one wanted to, find out, that therefore in principle there are no more mysterious incalculable powers that come into play here, that to the contrary one could, in principle, dominate all things through calculation.[6]

MATHEMATICS, SPACE, AND TECHNOLOGY

A common understanding that the world has become calculable would be an indication of how our being in the world is known to be encompassed by mathematics. One way this understanding has become specific is in the experience of traveling through space. There was a time when travel was travail, a rich and often painful experience that step by step has been alleviated by technology.

In August 1805, the Lewis and Clark expedition had reached the foothills of the Bitterroot Divide in what is now Montana and on the 27th was prepared to cross it on its way to the Pacific. Here is the September 3rd entry of the Lewis and Clark journals:

> With all our precautions the horses were very much injured in passing over the ridges and steep points of the hills; to add to the difficulty, at the distance of 11 miles the high mountains closed [in on] the creek, so that we were obliged to leave the creek to the right and cross the mountain [7,000 feet or more] abruptly. The ascent was here so steep that several of the horses slipped and hurt themselves; but at last we succeeded in crossing the

mountain, and camped on a small branch of [the north fork of] Fish creek. We had now made 14 mile, in a direction nearly north from the [Salmon] river; but this distance, though short, was very fatiguing, and rendered still more disagreeable by the rain, which began at three o'clock. At dusk it commenced snowing, and continued till the ground was covered to the depth of two inches, when it changed into sleet.[7]

Once they had reached the tributaries of the Columbia, traveling was by canoe and became easier and faster. Still, they did not reach the Pacific till the middle of November.

Half a century later, 1859–1862, John Mullan and his soldiers and workers built a road that went from the Missouri to the Columbia, starting from Fort Benton in Montana and ending in Walla Walla in Washington State, some three hundred miles east of the Pacific.[8] Now ordinary people, horses, mules, cattle, and wagons could cross the Bitterroot Range. Starting on the east side in Missoula in Montana it would take travelers three or four weeks rather than three months to reach the vicinity of the Pacific.

This crucial link of transcontinental travel and transport lost much of its importance when the Northern Pacific, the third transcontinental railroad, linked Missoula with Chicago to the east and Seattle to the west. The final golden spike was driven into its tie in 1883 at Independence Creek, fifty-nine miles up the Clark Fork from Missoula. The Milwaukee Road came through Missoula in 1909. Both lines in turn were eclipsed by the construction of Interstate 90 that completely and finally linked Missoula to Seattle in 1991. The Milwaukee Road is no more. The Northern Pacific surrendered its Missoula to Spokane portion to Montana Rail Link in 1987, which in turn became part of the Burlington Northern Santa Fe in 2022, and there is no more passenger traffic through Missoula. But who needs it? I-90 gets you from Missoula to Walla Walla in a comfortable five and a half hours. If you're in a hurry to get from Missoula to the Pacific, however, airlines will take you to Seattle in an hour and a half. The distance of five hundred miles or so that was hard and long for Lewis and Clark and their comrades and extended over months has softened and shrunk as technology has expanded.

The experience of space became a variable with technology the divisor in three senses of the word. Mathematically, as the divisor increases, the fraction of experience decreases and tends toward nothing as the divisor approaches infinity. In a rare and absolute sense, a divisor is an agent of division, and in this sense, technology is a device that divides ordinary life in an advanced technological society into the attractive surface of control and consumption and an underlying machinery that is concealed and to most people unintelligible. In Latin finally a *divisor* is a distributor of bribes. Modern technology as a way of dealing with reality is in many ways an admirable and beneficial enterprise, no doubt. Still, while it does not really bribe ordinary people into unsavory actions, it offers powerful inducements to detrimental behavior, to gaining weight, yielding to distraction, and lapsing into civic indifference.

MATHEMATICS, TECHNOLOGY, AND MEASUREMENT

There is yet another way that division unites mathematics and technology and that the joint force of the two informs the modern world and our being in the world. To measure is to divide a length of space (or time) into units. The oldest and simplest way to measure spatial distance was in terms of paces or a day's journeys. More formal measurements gave rise to geometry. They go back thousands of years. The builders of the Vedic fire altars in India must have known and applied the Pythagorean Theorem some three thousand years ago.[9] Meriwether Lewis took surprisingly accurate geographical measurements on the expedition from St. Louis to the Pacific Coast. The building of the railroads required punctilious surveys, schedules in standard time, and formally structured corporate organization.

Mathematics gained incomparably greater importance in the contemporary culture through information and communication technology (ICT). Mathematics now, rather than merely being applied to reality, inheres in reality. Mathematically, all ICT is a realization of Boolean algebra, which has two values, the ones and zeros everything comes down to, we are told. What makes those ones and zeros move in the intended way are the Boolean functions that are structured as logic gates and materialized in transistors.[10] The substructure of information theory and technology is for the most part screened off from our comprehension. We ordinarily touch only the input and output of its functions. I press the START button in my car and I hear the engine turn over; then I touch the symbols, letters, and numbers on the screen of the navigation system. Up comes the miles and the time to Helena, MT and my start on a moving map.

We all know that our power does not come from a genie that came out of a bottle. There are machineries under the hood and behind the screen; and, more important for my purposes, we know that these artifacts are pervaded and governed by immaterial structures—software, algorithms, programs, apps, operating systems, and the like. They are reliably rule-governed, and we are aware of the endpoint of those rules and connect with them by complying with instructions. We also have a sense of the extent and connections of these structures. My automotive navigator works because it is inserted into the omnipresent Global Positioning System (GPS). That system is in turn part of a global monitoring system that keeps track of military sites and movements and more mundanely of the earth's temperature, humidity, vegetation, and even of the movements of wolves and bears.

These animals with their electronic collars are an instance of all the objects and processes that leave electronic traces to be gathered into data and statistics. Phenomena that are too massive or diffuse to leave a helpful physical trace are captured (more or less) through surveys and polls. We reach for events that lie on the far side of the present by means of forecasts and futures markets. We measure the economy in terms of employment, investments, sales, and comprehensively through the gross domestic product and the standard of living. We measure people's body mass index, their fitness, and happiness. We reach into the future of sports and politics. There is, finally, the hope that in the vastness of big data, there already are previously elusive phenomena we can come to know if only we can uncover them. As Weber put it, we believe that "there are no more mysterious incalculable powers."[11]

Applied mathematics takes on a life of its own when it goes beyond taking measurements of phenomena that exist out there in the real world and models the essence of phenomena in mathematical structures. Models step forward as counterparts or in fact rivals of what's out there. Models of air flowing across the wing of an airplane tell us how to shape actual wings. Models of financial transactions try to stay ahead of reality to tell you how to manipulate or benefit from markets. Fantasy football departs from reality and returns to it, but in between there emerges what a participant considers the ideal invincible team. In contemporary video games the divisions of technology are triumphant. As a denominator, technology has become so expansive as to reduce the fraction of actual experience, i.e., the experience of the tangible world, close to zero.

A video game is a world unto itself. Its division from reality is total except for recollections of castles and forests, of persons and of beasts that are resources for the construction of a fantasy world. At the same time, the virtual reality of a game is governed by rules that are intricate and challenging and are engineered to bribe a player into staying in the game to the point of exhaustion. A video game, being a world unto itself, has its own space and time. Remarkably the game itself is positioned within a world that is well-structured and yet without measures of space and time. It's a realm of ubiquity and instantaneity. All the sites in cyberspace are equally near and far, and they come to be present in an instant. But they are connected to each other in an orderly fashion by way of linkage, inclusion, loops and hierarchies. Many sites *within* cyberspace have measures and distances of space and time. Video games certainly do. But the distancelessness of cyberspace to various degrees penetrates video games and radiates out into the actual world. The actual positions of participants in multiplayer games are immaterial. As for time, the great measure of the human condition, mortality, is a reversible feature within the game, not the end of the game; it's never true that *les jeux sont faits*.

Distanceless ubiquity and instantaneity, which are so evident in the structure of cyberspace, are penetrating the tangible world by degrees, penetration and prosperity moving in tandem. The progress of technology has transformed travel from travail into boredom, the impatient toleration of what rigidity has remained in the tangible world. The same impatience rules the ordinary experience of time. We expect food, entertainment, and news to be at the ready and, when beckoned, to show up in the blink of an eye; the more prosperous we are, the more impatient our expectations.

The divisions of technology have woven mathematical structures into the fabric of ordinary life. It's obvious enough that putting the presence of mathematics this way reveals troubling features of ordinary life. But there is also something characteristically contemporary and positive in the impact mathematics has had in ordinary life. The pervasive mathematical structures in everyday life can give us a sense of rational and universal order, the feeling that the messy life of ordinary existence is encompassed by and suffused with an ideal and abiding order that everyone, "if only one wanted to," as Weber said, can employ and apply to illuminate the world in a way that has a claim on everyone.

NOTES

1. George Lakoff and Mark Johnson, *Metaphors We Live By* (Chicago: University of Chicago Press, 1980).

2. Kent C. Bloomer and Charles W. Moore, *Body, Memory, and Architecture* (New Haven: Yale University Press, 1977), pp. 1–5.

3. See "Beauty is Suffering [Part 1: The Mathematician], https://www.youtube.com/watch?v=i0UTeQfnzfM, checked Jun 22, 2021.

4. Eugene P. Wigner, "The Unreasonable Effectiveness of Mathematics in the Natural Sciences," *Communications in Pure and Applied Mathematics*, volume13 (1960), pp. 1–14.

5. Thorstein Veblen, *The Theory of the Leisure Class* (New Brunswick: Transaction Publishers, 2000 [1899]), p. xx.

6. Max Weber, *Wissenschaft als Beruf* (Stuttgart: Reclam, 1995 [1917]), p. 19.

7. *The History of the Lewis and Clark Expedition*, ed. Elliott Coues (New York: Dover Publications, no date [1814ä volume 2, p. 580.

8. Ken Robison, "Mullan Expedition a Journey of Will," *Missoulian*, May 9, 2022, pp. A1 and A5.

9. See my *Holding On to Reality: The Nature of Information at the Turn of the Millennium* (Chicago: University of Chicago Press, 1999), pp. 65–68.

10. See my *Holding On to Reality*, pp. 125–65.

11. Weber, *Wissenschaft als Beruf*, p. 19 (my translation).

5

Ordinary Cosmology and Physics

"THE MOST FORTUNATE THOUGHT":
EINSTEIN AND GENERAL RELATIVITY

In 1920, when he was living and teaching in Berlin, Einstein put pen to paper and wrote:

> When (in 1907) I was occupied with a comprehensive paper on special relativity for the *Yearbook for Radioactivity and Electronics*, I had to try to modify Newton's theory of gravitation so that its laws would fit into the theory. Attempts undertaken in this direction showed, to be sure, the feasibility of this undertaking but did not satisfy me because they had to be supported with physically unfounded hypotheses. That is when the most fortunate thought of my life occurred to me in the following form:
>
> For any observer the gravitational field has only a relative existence just like the field that is generated by magnetic-electric induction. *This is because for an observer, freely falling off the roof of a house, there exists during his fall*—at least in his immediate environment—*no gravitational Field.* For if the observer lets go of whatever object, it stays with him in a state of rest, i.e., of uniform motion, independently of its chemical and physical composition. Hence the observer is justified in interpreting his state as "rest."[1]

It's fortunate for us that Einstein, though he was a fine mathematician, was prompted in his discoveries by his conceptual and intuitive imagination. The mathematical explication of general relativity turned out to be difficult, and it is not clear whether Einstein or David Hilbert first found the solution. Einstein himself stressed the priority of intuitive content over mathematics. At the start of the manuscript we're considering, Einstein indicates his pleasure at presenting his relativity theories to "nonmathematicians" and goes on to say:

I am doing this the more gladly as there is a certain danger that the unfortunately rather complicated form of the theory threatens to obscure its rather simple physical content. The mathematical form is merely a means whereas what is essential in the theory certainly consists in the fact that it is the consistent application of some simple principles. The theory is not at all the result of audacious speculations as is widely believed.[2]

Tejinder Knaur, David Blair, John Moschilla, Warren Stannard, and Marjan Zadnik realized that there is a way of modeling an instance of Einstein's scientific imagination. They used a membrane over a quadratic frame to show how, in the two-dimensional case, balls, say steel ball bearings, deform the flatness of the membrane into the curved space of Einstein's General Relativity and how spheres, say golf balls, roll down the membrane in ways that are analogous to the motions of bodies and photons in the curved space of General Relativity.[3]

The device Knaur and his coworkers constructed consisted of a frame of four ten-foot tent poles and a four-by-four Lycra membrane, quite a bit smaller than, and stretched across, the frame to assure appropriate elasticity of the membrane. The counterparts of the heavenly bodies were golf balls. This setting allowed Knaur et al. to model gravitational lensing. Its mathematical structure is represented by the equation:

$$\alpha = \frac{4GM}{c^2 b}$$

where α is the deflection angle of gravitational lensing, b is the impact parameter in astronomy, G is the gravitational constant, M is the Mass of a lensing object, and c is the speed of light. The impact parameter is a function of the distance of a star or galaxy from Earth—the farther the heavenly body from Earth, the larger the impact factor.[4] "Mathematically competent students," say Knaur et al., "can use this formula to estimate the mass of a lensing galaxy assuming an impact parameter of say 10^5 light years."[5] Given a mass at the center of the membrane, say ten steel ball bearings, that deforms the membrane, students can use pairs of toy cars to model the deflection behavior of photons in relation to the star or galaxy, represented by the mass at the center of the membrane.[6]

The membrane device can also be used to model good-old-fashioned Newtonian gravity. The force of toy cars with fixed steering, descending toward the center of the membrane whose force is measured with a spring balance and its metric represents the increases in a car's force at ever shorter distances from the center of the membrane. A plot of the results can be captured by the equation:

$$F = A \frac{m_1 m_2}{r^2}$$

The analogy to Newton's law of gravitation is evident:

$$F = G \frac{m_1 m_2}{r^2}$$

Thus, we can get a sense of the actual structure of space at its largest and of the paths celestial bodies take in curved space. That sense can become part of our background knowledge of being in the world.

THE QUANTUM WORLD

When you look at the presence of mathematics in ordinary cosmology, you can count on some degree of comprehension, on some low level of numeracy. Physical literacy, however, is lower yet. But if ordinary people have been unequal to literacy in physics, physics has been unequal to the task of providing a consistent and comprehensive theory of how the world is ultimately structured. There is the failure of our current physics to explain the large majority of what's out there in the world, the roughly 25 percent of the universe that consists of dark matter and the roughly 70 percent consisting of dark energy. Of the former we only know that it is subject to gravity and of the latter that it is kinetic energy, driving the universe apart at an accelerating rate. It's 5 percent of the world that quantum mechanics and relativity theory explain, with wonderful elegance and precision, to be sure. But these two great theories are inconsistent with one another.

There is, moreover, a crisis within our view of the cosmos at its largest. Both Boeke and the Morrisons mention the expansion of the Cosmos when they consider it at a remove of 10^{25} or 10^{26} meters.[7] They take expansion as well-established and fairly well-understood fact. But in 2020, Richard Panek pointed out "How a Dispute over a Single Number [the rate of expansion] Became a Cosmological Crisis."[8] Now what? "Experimental problems could cause the discrepancy, but no one is sure what those problems would be. Another possibility is that the conflict points to undiscovered phenomena—'a new physics.'"[9] Would a new physics have a bearing on moral cosmology? Yes, it might well, but it would not undermine it. A creditable moral cosmology would in any case be possible as I will suggest in the conclusion of my project.

Quantum physics, however, is troubled by its own and unique question: What does the world have to be like for quantum physics to be true? The behavior of reality at the quantum level seems so at odds with how manifest reality manifestly behaves that we may have to think of quantum mechanics not as a theory of reality, but merely as a device of computation and control. But we may ask, what then is reality at the smallest like? One answer, attributed to N. David Merman, is: "Shut up and compute." Some physics is being taught in high school, so ordinary cosmology has a foothold there. However, high school physics is mostly concerned with Newtonian mechanics which, to be sure, has a wonderful simplicity and a pleasant affinity with manifest reality. Regarding simplicity, you can bake Newtonian physics from three ingredients—space, time, and mass. Space (distance) divided by time (interval) and given a direction, gives you velocity; velocity multiplied by mass gives you momentum. Velocity divided by time gives you acceleration. And acceleration multiplied by mass gives you force. To this lawfulness you have to add the laws of inertia, reciprocity, and gravity. All this makes ready sense. We know that our velocity in running from home to the bus stop determines the time it takes to get from here to there—greater velocity, less time. We know that to accelerate on our way to the bus means that our velocity has to increase. We know that being bumped by a child coming at us at twenty miles per hour will give you the same shove as an adult, twice as heavy, coming at half the velocity; it's the same momentum. And perhaps we can say that the wind has real force when it hits you at a rising speed.

The supposed strangeness of quantum mechanics may be due in part to the fact that physicists are in the thrall of Newtonian simplicity and predictability. Yet some of the features of the quantum world have a natural affinity with the ordinary world and ordinary cosmology though they are at odds with the Newtonian world. The quantum world is inherently probabilistic, unlike the deterministic world of Newton. Uncertainty is congenial with our ordinary sense of being in the world. We do think of the future as ambiguous and of expectations as more or less likely to be fulfilled; and just as the probabilities of the quantum world are resolved through events (through measurements or decoherence), so events tell us that no, our candidate was not elected, or yes, you did get the job offer. We can in imagination connect in an instant to locations however distant, the Manoa Valley on Oahu in Hawaii or Alpha Centauri in our galaxy. Finally, we see our world branching all the time into the worlds of the roads not taken—an offer we declined, a move to Vermont not made, a death we did not expect. You may contemplate the world in which you would have become a professor in Chicago, the world in which you would have been an attorney in Vermont, or the world in which your mother would have seen your children. Thus, ordinary sensibilities have a significant kinship with the quantum world. But disclaimers are in order.

Quantum phenomena such as the fact that an electron is both a particle and a wave are strange, no matter what. The kinships that do exist are broad. The probabilities in quantum mechanics are lawful while those of the everyday world are contingent. The two kinds of probability reside on different levels of the foundational order, quantum probability at the very bottom, contingent probability at a high level of emergence. In ordinary experience, arbitrarily long distances and multiple worlds are imaginary. The distances that are crossed instantaneously (if "crossing" makes sense in this case) between two entangled particles are real. So are the many worlds in the many-worlds-interpretation of quantum mechanics if you accept this interpretation. And yet, these features at different levels can be part of the same attunement to the world.

SCIENCE AND TECHNOLOGY—HEIDEGGER

If modern technology has prompted implicit experiences of mathematics, might it not also have transmitted the structures of physics? A simple and misleading answer comes from philosophers who say that science and technology are at bottom just one phenomenon—technoscience. This view is shared by most of the media, *Scientific American* included. At most, however, technology and science are two aspects of one phenomenon, as the *New York Times* and Google News recognize. The technoscience view seems to be supported by the fact that science and technology, beginning in the middle of the nineteenth century, have become increasingly interdependent. Even at the very beginning of modern science at the turn from the sixteenth to the seventeenth century, Galileo used simple technologies to discover or confirm his theories of physics—telescopes, clocks, inclined planes, balls, etc. And today, finally, the majority of scientific research is undertaken to support and advance technologies of all kinds.

Martin Heidegger (1889–1976) is often thought to have identified science with technology. In *The Question Concerning Technology* he says:

> According to the chronology of history, the beginning of modern natural science lies in the 17th century. However, the motorized technology of machines does not develop until the second half of the 18th century. And yet what is later from the point of view of historical statement—modern technology—is historically earlier as regards its predominant essence.[10]

This was first published in 1954, first presented in 1949, and worked out in its essentials from 1934–1936. In 1935, Heidegger taught a course on *The Question of the Thing*.[11] That's where he discussed early modern science in a different vein and in a way that supports the need to take physics seriously as a theory that may bear on moral cosmology.

> The greatness and superiority of natural science in the 16th and 17th centuries are based on the fact that those researchers were philosophers one and all; they understood that there are no mere facts and that, to the contrary, a fact is what it is in light of a concept that justifies it and according to the scope of such justification.[12]

In "A Conversation of a Threesome on a Country Road" from the winter of 1944–1945, Heidegger acknowledges the view that science is discovery rather than invention, a view set forth by the researcher (the other members of the threesome being a sage or guide and a professor):

> Nature, and nature only as it reveals itself, has the last word in physics. It is among the overwhelming experiences of a natural scientist that nature often responds otherwise than the questions a researcher poses to it would lead one to expect.[13]

A partial resolution of these seemingly incompatible views can be found in an entry in the Black Notebooks of 1946–1947:

> There was a time when the sciences were released from philosophy to their autonomy which they were never to achieve, but were destined to surrender so that in their modern shape, seemingly free, they have yet become subservient to technology whence issues the nature of the sciences as well.[14]

The event that released the sciences from their tutelage of philosophy to their autonomy was the Enlightenment, the cultural beginning of the modern era. But is it true that, having escaped philosophy, the sciences surrendered to technology? To the extent they did, they helped to create a new kind of ethics, not the great modern ethics of political liberty and equality, but the ethics of liberation from the claims and burdens of reality. We have in chapter 4 seen an illustration of technological ethics in the development of travel and transport. It goes from the removal of dangers and miseries to the detachment from reality. If the sciences were inextricably enmeshed in technology, they would be bound to the morality of technology and unavailable as a force of moral cosmology.

Heidegger to the contrary, the sciences have kept their autonomy as the disclosers of the ultimate structures of reality. They have allowed nature to speak in newly fundamental and illuminating ways. It's at least possible that a moral cosmology emerges in part from what the sciences disclose. To be sure, in scientific research, technology is a crucial means of putting questions to science. But as Heidegger said in the winter of 1944–1945, "Nature, and nature only as it reveals itself, has the last word in physics." It is in physics where the purely instrumental role of technology and the purely disclosive power of the sciences are most evident. Technology as a cultural force lives and dies by its usefulness. But the exceedingly costly and sophisticated technologies that disclose reality at its smallest and largest, the Large Hadron Collider, e.g., and the Hubble Telescope, are useless to comfort and consumption. All they do is tell us what reality is ultimately and uselessly like.

POINTLESSNESS ABOVE AND BELOW TECHNOLOGY

In the advanced industrial countries, the culture of technology, it seems to me, took a largely negative turn around the middle of the twentieth century. Since then, common knowledge has declined, overweightness and obesity have increased, and the climate has begun to change noticeably and injuriously due to gigantic enterprises that have harmed the planet and done little for global justice. The culture of technology is a complex phenomenon, exhibiting a multiplicity of patterns. Some patterns trace the distribution of power and prosperity, others trace the transformation of the good life. The ones that concern us here are reflections of physics in ordinary life. The several patterns of technology are interwoven, of course, and the selection and consideration of a particular pattern must ultimately be moral and help us answer the question: What is the good life in a technological society?

The premise of this book is of course that a moral cosmology is needed for a satisfactory answer. If the premise is warranted, we should find reflections of physics in the ordinary experience of the world. Since physics is written in the language of mathematics, we find physics in ordinary cosmology wherever we find mathematics. Everything that is intuitively felt to be computable has physical dimensions attached to it—the metrics of time and space, the units of mass and energy, etc. Thus, the ordinary person knows, however dimly, that not only is mathematics in play when one uses a GPS device or plays a video game, but also the physics of rockets, satellites, and electronics, usually under the more general heading of "science." When someone says that understanding one's smart phone "is not rocket science," there's an acknowledgment that there is science in getting a GPS system aloft.

There are more deeply felt reflections of physics in the ordinary sense of being in the world. Although Newtonian physics had to yield to relativity and quantum theory, in engineering practice it has maintained its usefulness; and in the experience of ordinary cosmology, contemporary technology has created a Newtonian world of absolute time and space through the Global Positioning System (GPS). It assigns each of us its place on the planet and covers all of us with Universal Time that is translated for each of us into local time. This Newtonian world, however, has its basis in and is itself the basis for a pointless post-Newtonian reality. GPS relies on the electronics that quantum mechanics has made possible, and due to its relatively large distances and minute time intervals, it has to take account of spatial and temporal relations that are not fixed by an absolute framework but are relative to the velocities between the points of reference.

This post-Newtonian technology supports the quasi-Newtonian GPS framework of universal time and space which in turn supports the post-Newtonian ordinary cosmology. In the experience of advanced industrial societies, the rigid boundaries of time, space, and matter are beginning to dissolve. Artificial light has softened the lines between day and night; heating and cooling have reduced the contrasts of the seasons. These developments have preceded and opened the doors for the dissolution of the bounds that once had given habits and events their firm places.

Information and communication technology (ICT) makes definite places and times of work immaterial. Increasingly, you can work wherever and whenever. Events are detached from their unique times and places through recordings you can call up when and where you please. We even try to wrest events from the future through polling, projection, and prognosis. The demands of the here and now in social engagements are lifted through the asynchronous devices of voicemail, email, texting, and more generally through social media.

Thus, our world consists of three spheres or contexts that are nested in each other. The outermost shell is the universe. "The more the universe seems comprehensible, the more it also seems pointless," says Steven Weinberg.[15] It's pointless cosmologically in not having some point for its center or having an absolute framework that would give a particular point a unique set of coordinates. But what Weinberg actually means is that the physical universe reveals no moral point—we no longer live in the world in which "the heavens declared the glory of God" as Weinberg says elsewhere.[16] Should we conclude from this pointlessness that a person's ego is to be the center point of each person's moral and physical world as Kant proposed? But wouldn't chaos be the result if each of us would impose their unique order on a pointless universe?

Kant told us to accept only the moral law we give to ourselves. But we are to do so, Kant added, as rational beings, each endowed with one and the same rational faculty; hence we will give each to ourselves the same moral law. But when it came to orienting oneself in space, Kant showed how one person might well regard herself as the embodiment of a three-dimensional order—up and down, front and back, right and left. But what if each of us tried to force the world into his particular coordinate system? To answer the question, we have to move from the first context to the second, from the pointlessness of cosmic space to the local island of spatial orientation, to the Earth. It's been long known that we can capture the orientation that the Earth affords within the solar system in longitudes, latitudes, and altitudes, not easily until the eighteenth century as far as longitudes were concerned, but easily and often automatically today, thanks to the Global Positioning System.

NOTES

1. The manuscript of Einstein's summary is in the Pierpoint Library in New York City. Einstein never published the essay, but it's now available in German in *The Collected Papers of Albert Einstein, Volume 7, the Berlin Years: Writings, 1918–1921*, ed. Michel Janssen, Robert Schulmann, Joszef Illy, Christoph Lehner, and Diana Kormos Buchwald (Princeton: Princeton University Press, 2002), pp. 245–78. For the editorial particulars of the essay, see this volume, pp. 279–81. This part of Einstein's paper is on p. 265. The translation is mine.

The original has "hat an einem betracheten" which in the first place has a misspelling (it should read "betrachten"), and secondly does not make sense. From the context it seems clear that Einstein must have intended "für einen betrachtenden" or "für einen Betrachter." Joergen Veisdal, "Einstein and Hilbert's Race to Generalize Relativity," https://jorgenveisdal.medium.com/einstein-and-hilberts-race-to-generalize-relativity-6885f44e3cbe, consulted September 6, 2021.

2. *The Collected Papers of Albert Einstein*, p. 245. The translation is mine.

3. Tejinder, Knaur et al., "Teaching Einsteinian Physics at Schools: Part 1, https://iopscience-iop-org.weblib.lib.umt.edu:2443/article/10.1088/1361-6552/aa83e4/pdf, consulted October 25, 2021.

4. Ibid., p. 5.

5. Ibid., Figure 3 on p. 5.

6. Ibid., Figure 2 on p. 4.

7. Boeke, p. 32; the Morrisons, p. [20].

8. Richard Panek, "How a Dispute over a Single Number Became a Cosmological Crisis," *Scientific American*, March 1, 2020, pp. 30–37.

9. Panek, p. 32.

10. Heidegger, "Die Frage nach der Technik," in *Vorträge und Aufsätze* (Pfullingen: Neske, 1954), p. 30.

11. Heidegger, *Die Frage nach dem Ding* (Tübingen: Max Niemeyer, 1962).

12. Loc. cit., p. 51.

13. Heidegger, *Feldweg-Gespräche*, ed. Ingrid Schüßler (Frankfurt: Vittorio Klostermann, 1995), p. 17.

14. Heidegger, *Anmerkungen I–V (Schwarze Hefte 1942–1948)*, ed. Peter Trawny (Frankfurt: Vittorio Klostermann, 2015), p. 314.

15. See note 62. This refers to *The Cambridge Dictionary of Philosophy*—which shows how cosmology has fallen out of fashion in the discourse.

16. See note 63. Refers to the *Routledge Encyclopedia* entry, making the point about cosmology being a scientific enterprise.

Conclusion

Your background knowledge, as we have seen earlier, is a measure of how fully and how well you are inhabiting the world. A helpful way of advancing our understanding of background knowledge is to consider its injurious limitations and helpful liberations.

Let's begin with the spatial dimension. If you have never traveled beyond the borders of the state you live in, your knowledge of the diversity of cultures is parochial, and when we send our students to study for a year in a foreign country, they come back as different persons, more worldly, more certain of who they want to be. Or consider a couple who have recently immigrated to this country and in their determination to have their daughter become fully at home in the United States are silent about their homeland. Eventually their daughter will insist on knowing her ancestry and will travel to her parents' homeland.

An admirable case of the desire to deepen familiarity with the past is Steven Weinberg, who figured prominently in chapter 1 and is a widely read writer on the impact of science on contemporary culture. And yet in his late seventies, we have to assume, he made this decision:

> A while ago I decided that I needed to dig deeper, to learn more about an earlier era in the history of science, when the goals and standards of science had not yet taken their present shape.[1]

The result was *To Explain the World*, an illuminating history of the emergence of science, beginning with Ancient Greece. Not all historians welcomed Weinberg into their fold. I do.

Background knowledge varies in width and depth across the citizens of this country. Take climate change. There are still people who deny that climate change is caused by humans (is anthropogenic) and is accelerating. But most of us can't help being aware of the rising oceans, the more intense wildfires, and the more raging floods; and most of us accept the fact that these new and alarming phenomena are caused by climate change. Again, most of us know that climate change is primarily caused in turn by atmospheric carbon dioxide and that its increase has been caused by burning fossil fuels.

To gain a rudimentary understanding of how background knowledge works in everyday life, let's follow the chain of causes by which carbon dioxide is produced in combustion. Carbon is an element with an electron shell of four electrons surrounding its nucleus. Oxygen is an element that has an electron shell of two electrons. In combustion, carbon gets to share its four electrons with the two electrons each of two oxygen atoms. Voilà, one carbon atom combining with two oxygen atoms (carbon dioxide) combine into a carbon dioxide molecule—CO_2. How does this background knowledge contribute to our being in the world fully and well? Come home with me, and we'll see.

I've got a fire in my glass-door woodstove. There is pleasure in seeing and hearing the fire—strands of flames waving upward, logs of fir crackling. But there is pleasure too in knowing that the carbon of the wood is being efficiently consumed and, together with the air's oxygen, is being converted into carbon dioxide. There is also pain in knowing that this colorless and fragranceless gas is trapping heat and accelerating climate change, the pain being lessened by knowing that, had I left the wood to rot in the forest, or to be consumed by a wildfire, carbon dioxide would have been produced slowly or violently anyway. Late this fall or early winter, satellites will scan and help us measure the areas that have been burned in the Rockies, and there is pleasure in knowing that the clocks in the satellites would have given us misleading readings had they been synchronized at liftoff. In that case, the clocks, due to the effect of special relativity, would have run slow, as we have seen in the previous section.

Such cognitive pleasure comes not only with the equations of physics of the cosmos, but also with the contingency that, adapting words of Stephen Hawking, "breathes fire into the equations and makes a universe for them to describe."[2] It's the contingency of terrestrial conditions within the solar system, within our galaxy, within clusters of galaxies, within the cosmic microwave background that is left over from the hugely energetic, but utterly simple Big Bang whence the structure of the cosmos emerged unpredictably. And contingency emerges at all levels, including the nearly ungraspable contingency constantly manifesting itself at the quantum scale.

What is symmetry? Symmetry is for some object to look the same from all angles. A daisy's flower is symmetrical only if the view is constrained to two dimensions. That is, it looks the same no matter how you turn it on a two-dimensional plane. This is two-dimensional symmetry. How do we get from two-dimensional symmetry to three-dimensional symmetry? If you have three-dimensional symmetry, everything looks the same not just when viewed as a single plane but from every possible conception of space in every possible dimension. If three-dimensional space becomes the background knowledge for everyone's conception of space, there cannot be a norm or a center that can be defined and determined within that space. This lack of a norm or center leads to disorientation.

It is perhaps not surprising that the prevailing human condition is one of being lost in space and time. This is the consequence of the view from nowhere. It can lead to utter disorientation because it makes the world look the same from all angles. Yet within that three-dimensional space, humans can overcome the disorientation by discovering focal things. Focal things are not derivable or definable within three-dimensional space, but they must be compatible with three-dimensional space.

Out of the swirl of contingency, the universe has produced and still produces emergent forms, each articulated in a particular shape. Atoms and then molecules emerge from the expanding energy of the Big Bang. Hydrogen atoms coalesce into stars, which give birth to and are in turn birthed by galaxies. Planets and solar systems emerge from new stars; on Earth the living cell emerges from nonliving matter. Four billion years later, symbolic consciousness emerges as the human species evolves to take its awe-inspiring place in the universe. Grounding a moral cosmology in such knowledge of the emergence of novelty generates wonder and gratitude. Mary Evelyn Tucker eloquently describes this universal emergence of novelty:

> From a single point in space and a single point in time . . . a roaring force from one unknowable moment all the elements in the universe burst forth. The elements self-organized into nebulae and stars and then galaxies and planets; in a startling burst of creativity, their patterns unfolded again and again. From those elements' unfolding came cells and the beginnings of self-replication and self-complexifying, life and lives developing like a fugue, variety unfolding, complexity unfurling, until, with the evolution of human consciousness, the generative urgency of the universe created a way to turn and contemplate itself. We are beings in whom the universe shivers in wonder at itself.

That shiver is occasioned by focal things and the practices that make them actual for human beings. Here contingency gives way to something beyond itself. As Henry Bugbee said, there are certain things about which we cannot be mistaken. Focal occasions are focal. But they can only focus the cosmos for us if we view them through a conceptual lens that is compatible with the laws of physics. Focal things require constraints within three-dimensional space. They have to be small enough for humans to grasp them. Things that are vastly larger are beyond the grasp and celebration of humans. Once discovered, focal things are the endpoint of moral discourse—they are self-warranting.

To tell people that science must inform their background knowledge is not to blame them for not grasping the scientific details. Rather, it is to invite them out of that impoverishment of being-in-the-world and the accompanying disorientation into the depth of space and time given by an understanding of contemporary physics. It is to invite them into a deeper understanding of our place in the universe that can inform the focal occasions and practices that they already know to be valuable. We may not always be able to completely understand or explain the depth of the world gathered and disclosed in these focal occasions, but we can acknowledge it.

That acknowledgment can be greeted with either skepticism or celebration. To greet it with celebration deepening into wonder orients us and makes it possible for us to be at home in the universe.[3]

Gratitude follows from this acknowledgment—gratitude both for the rich complexity of our evolving universe as well as for the richness of focal occasions that ground us. An imperative of universality also follows—the need to open the eyes of others to what has been disclosed. This grateful and universalizing response is immune to corrosive skepticism.

Cosmic background knowledge comes to be focused on self-warranting events when we can say:

> There is no place I would rather be.
> There is no one I would rather be with.
> There is nothing I would rather do.
> And this I will remember well.

There are various kinds of focal things and the practices that make them actual—sitting around the dinner table with your beloved, making music with your friends, skiing cross country with your daughter. What good are the joys of the suffusion of cosmic background knowledge and of the epiphanies of a focal event? They are good for nothing, they are good in themselves. They generate the forcefield of insight and the mood of being in the right place. They are evidence that even in these deeply troubled times we can be at home in a moral cosmology.

NOTES

1. Weinberg, *To Explain the World* (New York: Harper Collins, 2015), ix.

2. Stephen Hawking, *A Brief History of Time* (New York: Bantam Books, 1990 [1988]), p. 174.

3. Mary Evelyn Tucker, "A Roaring Force from One Unknowable Moment." Interview with Kathleen Dean Moore, *Orion* 34, no. 3 (2015); cited in Kathleen Dean Moore, *Great Tide Rising: Toward Clarity and Moral Courage in a Time of Climate Change.* Berkeley, CA: Counterpoint, 2016, p. 36.

Afterword

Opening the Door to Quantum Mechanics and Relativity Theory

A NORMATIVE COSMOLOGY FROM THE GROUND UP

I have tried to sketch the appearance and disappearance of moral cosmology, of that inclusive understanding of the world where all that is known informs the conduct of life. I concluded with ordinary cosmology, the comprehension of the world that is normal in this country and in advanced industrial countries. Ordinary cosmology turned out to be confined and unsettled. This is the cosmology we have. Can we and should we have a cosmology that takes account of all that is known to be important today and where everything that is so known significantly informs daily life? That would be the ideal cosmology. Let me use, however, a more sober and robust expression for what I'm after—a normative cosmology. If it's conceivable, then, being a cosmology it's all-inclusive, and being normative, it implies norms that pervade and guide the daily course of life.

A normative cosmology has to be laid out in two stages. The first has to gesture at what it is important to know. The second has to show how such knowledge can be present and effective in our normal activities.* The first part will be a sketch of general education and move from the ground up. What is ground? It's the final layer that you arrive at when you look at the world ever more closely, when you descend Kees Boeke's ladder and end up at the quantum level that grounds everything that we know. Quantum mechanics and physics generally have, ever since Galileo, been grounded in mathematics. So, let's start with numbers. The kind we first become fa-

---

*This Afterword constitutes the "gesture" of the first part. The second part is presented above in the Conclusion to the book.

miliar with are the natural or counting numbers: 1, 2 ,3, ... These are the numbers
we teach a child. We have her count apples, say eight:

$$1 + 1 + 1 + 1 + 1 + 1 + 1 + 1 = 8$$

Then we tell her to take away three and count what's left:

$$8 - 3 = 5$$

Add one back to the five:

$$5 + 1 = 6$$

But what if we now ask her to take away eight from the six?

$$6 - 8 = ?$$

Mathematicians have again and again come upon problems like this. What to do?
One solution always is to outlaw an operation that our mathematics can't handle.
The other is to enlarge the mathematics that fails to deal with a problem surfacing
within the mathematics at hand. Here we go from the natural to the whole numbers,
the integers, that extend via zero into the negative numbers:

$$..., -2, -1, 0, 1, 2, 3, ...$$

Now we have a solution:

$$6 - 8 = -2$$

Both with natural and with whole numbers we can divide and multiply as well as
add and subtract. So we ask our toddler to divide the eight apples into groups of two
and see how many groups we get:

$$8 \div 2 = 4$$

But what if we ask her to divide eight apples into three equal groups without
remainder?

$$8 \div 3 = ?$$

She gets three groups of two each and is left with two apples. Obviously two apples
cannot be divided into three equal groups of whole apples. But if she breaks both
apples into three equal pieces, she gets six pieces; and if she adds two pieces to each

of the three groups, she gets three equal groups without remainders. To break is to fracture; to fracture a number is to reduce it to fractions:

$$8 \div 3 = 2\frac{2}{3}$$

2/3 is a fraction, a quotient or ratio of two whole numbers, two and three. An irrational number like pi cannot be expressed as a ratio of two whole numbers.

You may also show your daughter that she can have groups of numbers where the number of groups and the number of numbers in those groups are the same, two groups of apples of two apples each:

$$2 \times 2 = 4$$

Or three groups of three:

$$3 \times 3 = 9$$

2×2 can be written as 2^2 and 3×3 as 3^2. Numbers can be squared. The result of squaring a number is a square. Four and nine are squares. So is 25. What is the number that, when squared, yields 25? Or more briefly, what is the square root of 25?

$$\sqrt{25} = ?$$

The answer is five of course.

Squaring a number and then getting the square root of the resulting square gets you back to the original number:

$$5^2 = 25$$
$$\sqrt{25} = 5$$

Conventionally we do not use question marks in equations. Instead, we give the answer we seek a tentative name, namely x:

$$8 - 10 = x$$

Once we have the answer, we say so:

$$8 - 10 = x$$
$$x = -2$$

Or:

$$8 \div 3 = x$$
$$x = 2\frac{2}{3}$$

Squares have their most celebrated and useful place in the Pythagorean Theorem, considered by the physicist David Mermin to be the "most famous equation of all time."[1] The theorem says that the square on the longest side, say c, of a right-angled triangle is equal to the sum of the squares on the other two sides, say a and b:

$$a^2 + b^2 = c^2$$

Now what if some computation makes you end up with:

$$x^2 + 36 = 0$$
$$x^2 = -36$$
$$x = \sqrt{-36}$$
$$x = ?$$

What's the square root of –36? It can't be six, because $6^2 = 36$, not –36. Can it be –6? But –6 x –6 = 36 since two minus signs come to a plus sign in multiplication. So, there is really no solution to $\sqrt{-36}$. But we can imagine one. $\sqrt{-36}$ is an imaginary number while the natural, whole, and rational numbers are real numbers (as are the irrationals). But how do you connect the imaginary with the real? Here again boldness overcomes the seemingly intractable. The trick is first to reduce the imaginary to its minimum by factoring out $\sqrt{-1}$:

$$\sqrt{-36} = \sqrt{(36 \times -1)} = 6 \times \sqrt{-1}$$

and assigning it to the term i:

$$\sqrt{-1} = i$$

Finally, you treat i as one more constant in your equations. If you're lucky and get $i \times i$, the imaginary becomes real, if negative:

$$i \times i = i^2 = -1$$

Now we have a solution to:

$$x = \sqrt{-36}$$
$$x = \sqrt{36} \times \sqrt{-1}$$
$$x = \sqrt{36} \times i$$
$$x = 6i$$

(Or –6i since 36 is the square of both +6 and –6).

The integers contain the natural numbers; the rationals contain the integers. Together they are the reals. Is there a kind of number that contains both the reals and the imaginaries? Yes, the complex numbers:

$$a + bi$$

where a and b are real numbers and bi is an imaginary number. When $a = 0$ you are left with imaginary numbers; when $b = 0$, you are left with the real numbers. Now we can get ready to enter the world of quantum mechanics.

GETTING READY FOR TRAVELS IN THE QUANTUM WORLD

In the quantum world you find residents, travelers, and tourists. How long do you have to live in the quantum world to become a resident? "It takes seven years—four undergraduate and three graduate," says Richard Feynman, who knew what he was talking about.[2] What's a tourist in the quantum world? A person who gets on a trolley and sees the sights that a guide points out through the window. And a traveler? She asks for the trolley to stop and gets out into the quantum world, there to stay for a while. And what's the door that has to open and that she steps through? Equations. If you read a book about quantum physics that has no equations, you're staying in the trolley for a long, entertaining, and instructive time perhaps, but you haven't touched the ground of quantum physics. There are several excellent traveler's guides to quantum mechanics.[3] It takes months to follow them conscientiously. All we can do here is open the door of the trolley and take a few steps. We don't have a month or more.

Let's start with one of the affinities between daily life and quantum mechanics—uncertainty. Say there is an election. Your friend Alice is running against Bob. The polls give them equal chances. That state of affairs can be pictured on a Cartesian coordinate system with an arrow or vector. A vector, like an arrow, has a tip and a tail, a length, and a direction. Let's say its length is one unit, what of doesn't matter, an inch, a centimeter, whatever. We'll call the vector c. Let the tail be at the origin of the coordinates, and let the direction of c be 45° relative to the positive y-axis and to the positive x-axis.

We now know enough to figure out the coordinates of the tip of the vector c. Let's call the y-coordinate a and the x-coordinate b. The angle of c with a and of c with b is 45°. Thus, the triangle composed of a, b, and c is right-angled isosceles triangle where:

$$a = b$$

And the triangle, being right-angled, is subject to the Pythagorean Theorem: $a^2 + b^2 = c^2$ and since $c = 1$ and $a = b$ we get: $a^2 + a^2 = 1$ and $a^2 = \frac{1}{2}$ and $b^2 = \frac{1}{2}$ since $a = b$.

If the vector represents the state of the campaign between Alice and Bob and the vector is at forty-five degrees, the squares of a and b represent the odds of a (Alice) and of b (Bob), and at this moment the race is even between Alice and Bob. Chances of her victory are 50 percent, and so they are of his. The vector represents a state of uncertainty.

The sum of the probabilities of an event has to be one—certainty. The sum corresponds to the certainty that *a* or *b* will come to pass. Given the state of the campaign we're considering, one half and one half sum to one as they should. Is this just a happy coincidence or do the probabilities of the campaign always sum to one? If there is a happy coincidence, it is between mathematics and physics. The probabilities are always the squares of the two smaller sides (the "legs") of a right-angled triangle, of *a* and *b*. The sum of the squares of the legs has to be equal to the square of the hypotenuse. If the hypotenuse equals one, the square of the hypotenuse equals one and therefore

$$a^2 + b^2 = 1$$

whether *a* and *b* are equal or unequal. Assume for instance that there's been a change in the state of the campaign, reflected in the angle between *c* and *b* decreasing from 45 degrees to 30 degrees. What would that mean in the real world? The odds have shifted in Alice's favor. Again, geometry allows us to figure out the new probability of Alice's winning. The internal angles of the new triangle, composed of *a*, *b*, and *c* are 30°, 90°, and 60°.

b is evidently half of *c*, and the square of one half is one quarter or 0.25. These are the odds of Bob's campaign. They have decreased by half. Consequently, Alice's chances are now three quarters or 0.75. They have increased by half.

Imagine the vector continuing to pivot counterclockwise. As it pivots, *b* keeps shrinking till it reads zero; at that point, the vector *c* coincides with *a*, and hence: *a* = *c* = 1. The vote has taken place. Alice's victory is now certain; Bob's victory is no longer possible.

In real life, probability, its changes, and its resolution are not so neat. As we know, polls have been unreliable tools to determine probabilities. More significant for our purposes, probability as such is improbably puzzling. Consider the procedure that is often used to determine which of two possible events with equal probability is to become actual—flipping a coin. Why do we trust the procedure so often and implicitly? There is a mathematical or theoretical answer and an empirical or experimental answer.

Mathematically, you start with the assumption that in a particular case or device there is a number of equally possible outcomes, two in the case of a coin, six in the case of a die. Then the probability of a particular outcome is simply the reciprocal of the number of possible outcomes, 1/2 for a coin, 1/6 for a die. Thus the probability of heads is 1/2; the probability of a five coming up is 1/6. Now you can go to more interesting questions such as: What is the average outcome of flipping a coin or of rolling a die?

Here we come to the fork between the theoretical and the empirical way of demonstration. Theoretically or mathematically, there is a formula for the average of outcomes: Sum the possible outcomes, each multiplied by its probability. To apply the formula to a coin we have to assign numbers to heads and tails, say 1 for heads and −1 one for tails. Thus, the average of outcomes is

$$1 \times \frac{1}{2} + (-1) \times \frac{1}{2} = 0$$

Dice already have numbers on their faces. Thus, the average of the six outcomes is

$$1 \times \frac{1}{6} + 2 \times \frac{1}{6} + 3 \times \frac{1}{6} + 4 \times \frac{1}{6} + 5 \times \frac{1}{6} + 6 \times \frac{1}{6} = 3.5$$

In the case of a coin, the empirical, experimental, or observational method is to flip a coin again and again and add the outcomes one after another, a 1 for heads, a −1 for tails. The sum should be zero if the theoretical and the empirical methods agree. (Here you can ignore the 0.5 probability factor.) Do they? Well, you can see for yourself—flip a coin and do the additions. If you do, you're assuming that the coin is unbiased or fair and that your flipping is too, it's random. Your assumption is commonly shared, but it may give you pause. Even when the coin is fair, how come a deterministic creature like you in a deterministic world like ours can generate random events?

The results may at first tell you that you cannot. The first eight outcomes may be all ones—no minus ones. So, the sum is eight rather than zero as the theory would predict. If you persist, the sum will fluctuate above and below zero, but in the long run approximate zero ever more closely. This phenomenon reflects "the law of large numbers." Theorists differ on whether the emerging agreement of theory and empirical reality should give you pause. It gives me pause, but I accept the conclusion that theory and experiment converge in the long run.

OPENING THE DOOR TO THE QUANTUM WORLD

The theory that governs the world at large is relativity theory. It rules at the top rungs of the Kees Boeke ladder. Relativity theory was the discovery of one person, Albert Einstein, and the birth of its two stages, special and general relativity, were precisely marked by two papers published in 1905 and 1916.[4] Pencil and paper were the primary tools. Experiments played a minor role though they supplied sensational support once the theory had been published.

In 1921 Otto Stern and Walther Gerlach built a machine that produced unexpected and puzzling results on a screen. We'll set aside the details of the machine and the questions Stern and Gerlach had hoped to answer via the machine.[5] For our purposes we can think of the machine as spraying silver atoms through a vertical slit at a screen. We (and Stern and Gerlach) would expect the screen to show a fuzzy vertical line where the atoms hit the screen and left their marks. Instead, the screen showed two distinct clumps, one up and one down. To get the full force of what's so surprising and inexplicable in the Stern-Gerlach results, imagine we can intercept the beam that produced the up clump and block or ignore the down clump and send the up beam through a horizontal slit and onto a screen. The result would show a left clump and a right clump of marks. Finally, let's intercept the beam that produced the left clump, ignoring the right clump, and send it through another vertical slit. The result once again would be an up clump and a down clump on a screen.

What's strange about all this? Is it any different from sorting a flock of sheep into black and white sheep or ewes and rams? Up to a point it is not. Say you have a flock of sheep in a pen, and at the gate you have an up chute and a down chute. You send all the white sheep to the upper level and all the black sheep to the lower level. On the upper level, you have a left chute and a right chute. You send all the ewes through the left chute into the left pen, all the rams through the right chute into the right pen. So far so good. But just to make sure you made no mistake in your first splitting of the flock, you once more use an up and down chute on the upper left pen, again up for white and down for black. Since in the upper left pen you have only white ewes, you'd expect all of these white sheep to go up the up chute and none down the down chute. But not so—half of the ewes turn out to be white and go up, but the other half goes down; they're black. How can a totally white flock divide into white and black sheep?

Is such a thing unimaginable in everyday life? Consider this situation. You have a representative group of Americans. They're looking at an up ramp and a down ramp. You invite all the Democrats to go up and all the Republicans to go down. At the end of the up ramp is a fork, one going left, the other right. You invite all the believers to go right and all the nonbelievers to go left. At the end of the left ramp, on the upper left platform, there is another pair of up and down ramps. Once more, just to be sure, you invite the Democrats to go up and all the Republicans, if any, to go down. Half of the people go up, half go down. How can this be? There were only Democrats on the upper left platform. But perhaps half of these Democrats remembered, while taking the left fork, that they had been "mugged by reality," something that supposedly turns liberals into conservatives. Does everyday reality then furnish a plausible illustration of a quantum event that seems to violate common sense?

It does not, it only supports the need to transform common sense into a novel intuition. We have an explanation for the unexpected change of half the Democrats into Republicans. In quantum mechanics, all attempts to find an explanation have definitively failed. The change from all up into half up and half down is rock-bottom reality. There is no underlying explanation.

Here we run into a quantum fact that we can't fit into the ordinary ways of the everyday world. Is the quantum world then unintelligible? Is it impossible to understand it? Let me quote the answer Richard Feynman gave. He, if anyone, should have understood quantum physics, and yet he said what perhaps I should say at this point:

> Why are you going to sit here all this time, when you won't be able to understand what I am going to say? It is my task to convince you not to turn away because you don't understand it. You see, my physics students don't understand it either. That is because I don't understand it. Nobody does.[6]

Feynman, I think, spoke from unrequited love. He loved quantum physics and was more intimate with it than anyone else with the exception of Paul Dirac. Feynman had hoped to reconceive quantum physics. But when, as Frank Wilczek tells us, "He realized that his theory of photons and electrons is mathematically equivalent to the usual theory, it crushed his deepest hope."[7]

Still, I believe you can travel and reside in the quantum world and become used to that world. As with numbers, you begin with what's natural, and step by step you move into more abstract regions. You start with a campaign, picture it as a vector in two dimensions, go on to vectors in three-dimensional space and thence to the vectors in the infinitely dimensional Hilbert Space. Leonard Susskind, another eminent quantum physicist, wants you to forswear your ordinary intuitions about vectors: "Such vectors have three components, corresponding to the three dimensions of space. I want you to completely forget about that concept of a vector."[8] I'd like to suggest instead that you extend your concept and at least imagine what it would be like to be at home in that more extended world. Elsewhere Susskind appears to agree: "And eventually we do develop new kinds of intuitions."[9] So I will have you envision campaigns and sheep and then invite you to take steps beyond them into the farther reaches of the quantum world.

The puzzle that Stern and Gerlach discovered in their experiments led to steps in all kinds of directions. There was a series of puzzlements, of unwitting discoveries, of chance encounters, and of lucky guesses. These wanderings reached a first place of rest when George Uhlenbeck and Samuel Goudsmit met in 1925 and, combining their several talents and experiences, came up with the property that atoms and atomic particles possess and that allows you to give a consistent account of the Stern-Gerlach experiments.[10] That property is spin. Silver atoms, or their electrons, rotate, they spin. Spin found its most basic and elegant place in Paul Dirac's unification of quantum mechanics and special relativity of 1928. General relativity and quantum mechanics—quantum gravity—are still awaiting their reconciliation.

STEPPING INTO THE QUANTUM WORLD

Since the universe is a book that's written in the language of mathematics, as Galileo said, to see how the universe reads at bottom, we must render its structure in mathematics. The chunk we're trying to understand is its simplest part, the spin of an elementary particle or atom. Spin, like a campaign, has its states and probabilities. A particular spin is always in a certain state and, most simply, is rendered as a vector. In Dirac's notation it looks like this: $|A\rangle$

Think of it as an arrow if you like. Spin is something you can measure. What you measure is its direction. Like a campaign before the vote, spin, before it's measured, is uncertain. And just as the vote gives us a conclusive outcome, so does the measurement of a spin. The vote told us that Alice had won. A measurement tells us that spin is up, for instance. You would normally assume that if two spins differ, one might spin clockwise, the other counterclockwise. Clench your left hand and extend your thumb and have it point up. Your fingers will point clockwise, the thumb points up. If you turn your left hand upside down, the thumb points down, the fingers point counterclockwise. In quantum mechanics, the direction of spin is in terms of up, down, left, right, and whatever direction, rather than in terms of clockwise or counterclockwise.

A spin has components just as a campaign (*c*) has its components, Alice (*a*) and Bob or (*b*). The campaign is represented by a vector. We set its length at one, and when Alice and Bob had equal chances, the direction of the vector *c* was at 45 degrees to the x-axis and the y-axis. You can think of *a* and *b* as vectors as well. Both have a length and a direction. When Alice's and Bob's chances were even, the length of *a* was (1/√2), and the length of *b* was (1/√2) as well. What about the direction of *a* and *b*? *a*'s is in the up direction or along the y-axis, and *b*'s direction is to the right or along the x-axis. Hence, we can define the sum of *a* and *b* thus:

$$c = \frac{1}{\sqrt{2}}y + \frac{1}{\sqrt{2}}x$$

It's a sum that cannot be summed into one term. You cannot add x's to y's, only x's to x's and y's to y's.

The representations of the campaign, of Alice, and of Bob live in Cartesian space. Not so spins. They live in a space called Hilbert space. David Hilbert was a great mathematician who lived from 1862 to 1943 and who, as we see in chapter 5, competed with Einstein in trying to reach the mathematical explication of General Relativity. The Cartesian space we have been considering is two-dimensional. But it could be three-dimensional just like the space that Kant considered in his discussion of orientation in space and like the space that is usual in physics and in Hilbert space: the x-axis (left and right), the y-axis (in and out), and the z-axis (up and down). On each axis, we can imagine two vectors extending in opposite directions from the origin, i.e., from the intersection of the three axes.

Just as in two-dimensional Cartesian space you can define any vector as the sum of two vectors, in Hilbert space you can define any vector as the sum of an up-vector and a down-vector. In particular, you can define right and left and in and out vectors in terms of up and down vectors:

$$|r\rangle = \frac{1}{\sqrt{2}}|u\rangle + \frac{1}{\sqrt{2}}|d\rangle$$

To distinguish right from left, $(1/\sqrt{2})|d\rangle$ is negative in the definition of left:

$$|l\rangle = \frac{1}{\sqrt{2}}|u\rangle - \frac{1}{\sqrt{2}}|d\rangle$$

To distinguish in and out from right and left, the number that specifies $|d\rangle$ is $(i/\sqrt{2})$; *i*, as you will remember, is the minimal and residual number that simplifies our dealings with complex numbers. And as you may remember, a complex number is the sum of a real number, e.g., five and an imaginary number, which itself is the product of a real number, e.g., 3 and *i*: 3*i*.

Thus, the definition of in and out:

$$|i\rangle = \frac{1}{\sqrt{2}}|u\rangle + \frac{i}{\sqrt{2}}|d\rangle$$
$$|o\rangle = \frac{1}{\sqrt{2}}|u\rangle - \frac{i}{\sqrt{2}}|d\rangle$$

Note that $|i\rangle$ and i are not the same. $|i\rangle$ is a vector; i is ($\sqrt{-1}$).

Here's an indication why Hilbert space is complex (while Cartesian space is real)—it must have space for complex numbers which include the real numbers when the number that is multiplied by i is zero:

$$5 + (0 \times i) = 5$$

In quantum mechanics, a state of uncertainty is called "superposition." Thus $|r\rangle$ is a superposition of $|u\rangle$ and $|d\rangle$, and here too we get the probabilities by squaring the coefficients.

To give a mathematical account of the Stern-Gerlach experiments, we have to render the Stern-Gerlach machine, the states of the spin that the machine operates on, and the results of the measurements in numbers. We won't need complex numbers since we can ignore the $|i\rangle$ and $|o\rangle$ vectors. Integers will do for $|u\rangle$, $|d\rangle$, $|l\rangle$, and $|r\rangle$.

To begin with the vectors, they can be rendered in columns:

$$|u\rangle = \binom{1}{0}$$
$$|d\rangle = \binom{0}{1}$$

$|u\rangle$ and $|d\rangle$ are vectors of unit length along the z-axis. You can think of the vector $\langle 1,0 \rangle$ as extending one unit up from the origin of the coordinate of the z-axis and zero units on the down extension of the z-axis. The column vector for $|d\rangle$ is zero units on the up side of the z-axis and one unit down on the z-axis. You get the column vector for $|r\rangle$ by plugging in $\langle 1,0 \rangle$ for $|u\rangle$ and $\langle 0,1 \rangle$ for $|d\rangle$ in:

$$|r\rangle = \frac{1}{\sqrt{2}}|u\rangle + \frac{1}{\sqrt{2}}|d\rangle$$

and doing the math. What you get is:

$$|r\rangle = \binom{\frac{1}{\sqrt{2}}}{\frac{1}{\sqrt{2}}}$$

When you plug in the column vectors for $|u\rangle$ and $|d\rangle$ in

$$|l\rangle = \frac{1}{\sqrt{2}}|u\rangle - \frac{1}{\sqrt{2}}|d\rangle$$

what you get is:

$$|l\rangle = \begin{pmatrix} \frac{1}{\sqrt{2}} \\ -\frac{1}{\sqrt{2}} \end{pmatrix}$$

One of the relevant parts of the Stern-Gerlach machine is the slit that operates on the spray of silver atoms. It is rendered as a two-dimensional array of numbers, a matrix so-called. We'll call it M (also called *operator*). There's one rendition of the slit when it's in the vertical position, M_V, and another for the horizontal position, M_H:

$$M_V = \begin{pmatrix} 1 & 0 \\ 0 & -1 \end{pmatrix}$$

$$M_H = \begin{pmatrix} 0 & 1 \\ 1 & 0 \end{pmatrix}$$

How do machines operate on vectors? The general scheme is this:

$$\begin{pmatrix} m11 & m12 \\ m21 & m22 \end{pmatrix} \times \begin{pmatrix} a \\ b \end{pmatrix} = \begin{pmatrix} m11 \times a + m12 \times b \\ m21 \times a + m22 \times b \end{pmatrix}$$

Finally, let's agree that the result of measuring the direction of the spin and finding it to be up will be 1 and of finding it to be down to be −1, just as in talking about a coin we assigned 1 to heads and −1 to tails. We now have everything that's needed to render mathematically what happens in the Stern-Gerlach experiments.

THEORY AND EXPERIMENTS

Stern and Gerlach proceeded experimentally or empirically. They sent, I assume, trillions of silver atoms through the slit of their machine and onto a screen, and their results clearly met the requirements of large numbers. Our goal is to proceed mathematically or theoretically and assume it's a spray of electrons and their possible spins that is shot through the slit and hits the screen.

In the first case that we're considering the spray of electrons is sent through a vertical slit. Before they hit the screen, the electrons will be in a superposition of infinitely many directions, among them the superposition of up and down. Let $|A\rangle$ represent the spray of electrons in superposition. The mathematical representation of $|A\rangle$ and the superposition of up and down is:

$$|A\rangle = \frac{1}{\sqrt{2}}|u\rangle + \frac{1}{\sqrt{2}}|d\rangle$$

The effect of the slit on the electrons is represented by the application of M_V on the electrons in superposition:

$$M_V|A\rangle = M_V \left(\frac{1}{\sqrt{2}}|u\rangle + \frac{1}{\sqrt{2}}|d\rangle \right)$$

$$= \frac{1}{\sqrt{2}} M_V|u\rangle + \frac{1}{\sqrt{2}} M_V|d\rangle$$

$$= \frac{1}{\sqrt{2}} \begin{pmatrix} 1 & 0 \\ 0 & -1 \end{pmatrix} \times \begin{pmatrix} 1 \\ 0 \end{pmatrix} + \frac{1}{\sqrt{2}} \begin{pmatrix} 1 & 0 \\ 0 & -1 \end{pmatrix} \times \begin{pmatrix} 0 \\ 1 \end{pmatrix}$$

$$= \frac{1}{\sqrt{2}} \begin{pmatrix} 1 \times 1 + 0 \times 0 \\ 0 \times 1 + (-1) \times 0 \end{pmatrix} + \frac{1}{\sqrt{2}} \begin{pmatrix} 1 \times 0 + 0 \times 1 \\ 0 \times 0 + (-1) \times 1 \end{pmatrix}$$

$$= \frac{1}{\sqrt{2}} \begin{pmatrix} 1 \\ 0 \end{pmatrix} + \frac{1}{\sqrt{2}} \begin{pmatrix} 0 \\ -1 \end{pmatrix}$$

We can (trivially) resolve $\langle 1,0 \rangle$ into the factors 1 and $\langle 1,0 \rangle$, and we can resolve $\langle 0,-1 \rangle$ into -1 and $\langle 0,1 \rangle$. Hence:

$$M_V|A\rangle = \frac{1}{\sqrt{2}} \times 1 \begin{pmatrix} 1 \\ 0 \end{pmatrix} + \frac{1}{\sqrt{2}} \times -1 \begin{pmatrix} 0 \\ 1 \end{pmatrix}$$

$$= \frac{1}{\sqrt{2}} \times 1|u\rangle + \frac{1}{\sqrt{2}} \times (-1)|d\rangle$$

$(1/\sqrt{2})$ is the square root of getting the result $1|u\rangle$, i.e., that the electrons with spin up will show on the up side of the screen. As always, you square the coefficient $(1/\sqrt{2})$ to get the probability, one half in this case, i.e., 0.5; and likewise for the result of the spin showing on the down side of the screen. But if there is in fact an actual experiment what, according to the theory, will the direction of the spin turn out to be, up or down? Prior to experiment, no one can tell, no matter how precise the theory. Here we run into the aboriginal uncertainty of the quantum world that can only be resolved by an actual experiment, just as the outcome of a coin toss can only, and only unpredictably, be determined by actually tossing a coin.

Let's assume we have a spray of electrons coming through a vertical slit and we block the down beam of electrons before they hit the screen and that we intercept the up beam of electrons and send them through one more vertical slit before they hit the screen. What does the theory say about the outcome? Here's the answer:

$$M_V|u\rangle = \begin{pmatrix} 1 & 0 \\ 0 & -1 \end{pmatrix} \times \begin{pmatrix} 1 \\ 0 \end{pmatrix} = 1 \begin{pmatrix} 1 \\ 0 \end{pmatrix}$$

This agrees with experiments and says that once the up direction of the previously indeterminate spin has been determined, it remains up and can be unambiguously confirmed by measurement. It's like confirming that white sheep are white.

But what will happen according to the theory if we intercept the up beam that comes out of the vertical slit and send it through a horizontal slit represented by M_H? The electrons with spin up before they hit the horizontal slit will again be in a superposition of infinitely many directions, among them the superposition of right and left:

$$|u\rangle = \frac{1}{\sqrt{2}}|r\rangle + \frac{1}{\sqrt{2}}|l\rangle$$

With a little bit of algebra, you can derive this equation from two equations we have seen before:

$$|r\rangle = \frac{1}{\sqrt{2}}|u\rangle + \frac{1}{\sqrt{2}}|d\rangle$$
$$|l\rangle = \frac{1}{\sqrt{2}}|u\rangle - \frac{1}{\sqrt{2}}|d\rangle$$

Applying the horizontal slit to this state of the spin, we get:

$$M_H|u\rangle = M_H(\frac{1}{\sqrt{2}})$$

$$= \frac{1}{\sqrt{2}}\begin{pmatrix} 0 & 1 \\ 1 & 0 \end{pmatrix} \times \begin{pmatrix} \frac{1}{\sqrt{2}} \\ \frac{1}{\sqrt{2}} \end{pmatrix} + \frac{1}{\sqrt{2}}\begin{pmatrix} 0 & 1 \\ 1 & 0 \end{pmatrix} \times \begin{pmatrix} \frac{1}{\sqrt{2}} \\ -\frac{1}{\sqrt{2}} \end{pmatrix}$$

$$= \frac{1}{\sqrt{2}}\begin{pmatrix} 0 \times \frac{1}{\sqrt{2}} + 1 \times \frac{1}{\sqrt{2}} \\ 1 \times \frac{1}{\sqrt{2}} + 0 \times \frac{1}{\sqrt{2}} \end{pmatrix} + \frac{1}{\sqrt{2}}\begin{pmatrix} 0 \times \frac{1}{\sqrt{2}} + 1 \times (-\frac{1}{\sqrt{2}}) \\ 1 \times \frac{1}{\sqrt{2}} + 0 \times (-\frac{1}{\sqrt{2}}) \end{pmatrix}$$

$$= \frac{1}{\sqrt{2}}\begin{pmatrix} \frac{1}{\sqrt{2}} \\ \frac{1}{\sqrt{2}} \end{pmatrix} + \frac{1}{\sqrt{2}}\begin{pmatrix} \frac{-1}{\sqrt{2}} \\ \frac{1}{\sqrt{2}} \end{pmatrix}$$

$$= \frac{1}{\sqrt{2}}(1\begin{pmatrix} \frac{1}{\sqrt{2}} \\ \frac{1}{\sqrt{2}} \end{pmatrix}) + \frac{1}{\sqrt{2}}(-1\begin{pmatrix} \frac{1}{\sqrt{2}} \\ \frac{-1}{\sqrt{2}} \end{pmatrix})$$

$$= \frac{1}{\sqrt{2}}(1|r\rangle) + \frac{1}{\sqrt{2}}(-1|l\rangle)$$

That electrons with spin up should divide into those with right spin and those with left spin is no more implausible than a flock of white sheep dividing into ewes and rams.

Finally, let's assume we want to confirm that the electrons that had spin up and then were shown to also to have spin right still have spin up. A spin right, as we have seen above, is defined as the sum of an up spin and a down spin:

$$|r\rangle = \frac{1}{\sqrt{2}}|u\rangle + \frac{1}{\sqrt{2}}|d\rangle$$

Applying then the vertical slit to spin right, we get:

$$M_V|r\rangle = \frac{1}{\sqrt{2}}|u\rangle + \frac{1}{\sqrt{2}}|d\rangle$$

This result is analogous to the result of the first experiment, viz.

$$M_V|A\rangle = \frac{1}{\sqrt{2}}(1)|u\rangle + \frac{1}{\sqrt{2}}(-1)|d\rangle$$

One in this case means *right* and *minus one* means *left*. This result too agrees with experiment.

What if we want to reassure ourselves that the up component that the horizontal spin has inherited from the prior experiment is still up? Assume our latest result has been $1|r\rangle$. To check on the vertical component of $|r\rangle$, we apply the vertical slit M_V to $|r\rangle$:

$$M_V|r\rangle = M_V\left(\frac{1}{\sqrt{2}}(1)|u\rangle + \frac{1}{\sqrt{2}}(-1)|d\rangle\right)$$

It's obvious that the outcome of the computation will be the same as that of the first computation:

$$M_V|r\rangle = \frac{1}{\sqrt{2}}(1)\begin{pmatrix}1\\0\end{pmatrix} + \frac{1}{\sqrt{2}}(-1)\begin{pmatrix}0\\1\end{pmatrix}$$
$$= \frac{1}{\sqrt{2}}(1)|u\rangle + \frac{1}{\sqrt{2}}(-1)|d\rangle$$

The result is as strange as testing white ewes for blackness and whiteness and finding not just white ewes, but white *and* black ewes. This strange result can be formally captured in different ways.[11]

Having electrons pass through a slit and hit a screen is making a measurement (of the direction of spin in our case). Today there are machines that can send just one electron in superposition through a vertical slit. Let $|e\rangle$ represent such an electron that's in superposition of up and down:

$$|e\rangle = \frac{1}{\sqrt{2}}|u\rangle + \frac{1}{\sqrt{2}}|d\rangle$$

And let this be the representation of $|e\rangle$ being sent through a vertical slit:

$$M_V|e\rangle = M_V\left(\frac{1}{\sqrt{2}}|u\rangle + \frac{1}{\sqrt{2}}|d\rangle\right)$$

The result, as far as the theory will take us, is this:

$$M_V|e\rangle = \frac{1}{\sqrt{2}}(1)|u\rangle + \frac{1}{\sqrt{2}}(-1)|d\rangle$$

An actual experiment is as likely to show the electron up on the screen as it is likely to show it down on the screen. Let's say we do the experiment, and the result shows the mark of the electron up on the screen, there for all to see. We have done the experiment, and that experiment was an event. Did it take place in the quantum world or in the everyday world? For half a century the prevailing answer was the one that Niels Bohr and Werner Heisenberg had worked out in Copenhagen from 1927 to 1929. It said that measurement is an intervention in the quantum world carried out by an observer with an instrument, situated in the everyday world. The intervention makes the superposition of the quantum world collapse into a definite position in the everyday world, the superposition of a spinning electron showing up on a screen as up.

And what happened to the *down* vector in this case? That is one troubling question for the Copenhagen interpretation. Another and even more troubling question is: Are there two worlds? A quantum world and an everyday world (often called the classical or Newtonian world)? And if so how are they related? What, from the standpoint of physics, is it for one to intervene in the other? There have been many answers.

In 1957, Hugh Everett gave radical answers to these questions.[12] To the first question his answer was: If an experiment resolves the superposition into a definite up, the down spin does not disappear but shows definitely down in another world. And to the second question he replied: When a measurement occurs, there is only one world, the quantum world, and it contains not only the superposition, but the observer and the instrument as well. Let's put this in quantum notation. There is the spin in superposition of up and down:

$$\frac{1}{\sqrt{2}}|u\rangle + \frac{1}{\sqrt{2}}|d\rangle$$

We take the vertical slit M_V to be a quantum system as well:

$$|M_V\rangle$$

Let Otto Stern and Walther Gerlach constitute a quantum system and are represented by a vector:

$$|SG\rangle$$

Next think of multiplication as interaction. Then we get:

$$\left(\frac{1}{\sqrt{2}}|u\rangle + \frac{1}{\sqrt{2}}|d\rangle\right) \times |M_V\rangle \times |SG\rangle$$

And you apply the magic of algebra:

$$\left(\frac{1}{\sqrt{2}}|u\rangle \times |M_V\rangle \times |SG\rangle\right) + \left(\frac{1}{\sqrt{2}}|d\rangle \times |M_V\rangle \times |SG\rangle\right)$$

You get two systems of interacting parts connected by a plus sign. Just as in

$$\frac{1}{\sqrt{2}}|u\rangle + \frac{1}{\sqrt{2}}|d\rangle$$

up and *down* are totally distinct and won't interact in any way, so

$$\frac{1}{\sqrt{2}}|u\rangle \times |M_V\rangle \times |SG\rangle$$

and

$$\frac{1}{\sqrt{2}}|d\rangle \times |M_V\rangle \times |SG\rangle$$

are totally distinct and won't interact in any way. They belong to two different worlds. The one world prior to the measurement event has split into two separate worlds:

$$\left(\frac{1}{\sqrt{2}}|u\rangle \times |M_V\rangle \times |SG\rangle\right) + \left(\frac{1}{\sqrt{2}}|d\rangle \times |M_V\rangle \times |SG\rangle\right)$$

Everett's "many worlds" view has several virtues. It is deterministic. All possibilities are destined to be actualized. Everett's version has a second virtue. It is monist in the sense that, unlike the Copenhagen interpretation, it has only one kind of reality. But it fails to account for one feature in quantum mechanics. Assume for the sake of the argument that the one and only electron of a hydrogen atom is in superposition of two energy states, the ground state, $|G\rangle$, and the excited state $|E\rangle$; and let's further assume that the square root of the probability of it being in the ground state is $(\sqrt{3})/2$ and of it being in the excited state is $\frac{1}{2}$. Finally assume that you, $|Y\rangle$, have a device, $|b\rangle$, that allows you to determine the energy state of the hydrogen atom. Then we end up with this Everett version of the result of our measurement:

$$\left(\frac{\sqrt{3}}{2}|G\rangle \times |D\rangle \times |Y\rangle\right) + \left(\frac{1}{2}|E\rangle \times |D\rangle \times |Y\rangle\right)$$

Again, we have two distinct worlds where in one world the electron is in the ground state and in the other world it is in the excited state. Both of the possible worlds are destined to become actual. But doesn't that mean they are equally likely to become actual? So, what is the significance and what is the effect of the different coefficients, viz., $(\sqrt{3})/2$ and $1/2$, and therefore of the probabilities, viz., $3/4$ and $1/4$, of the two energy states?

Perhaps there are versions of the many-worlds interpretation that make sense of probabilities and of unequal probabilities in particular. That would still leave us with a consequence of the many-worlds interpretation that many quantum theorists find unacceptable. Measurements carried out by humans are not the only events of resolving superposition in the many-worlds interpretation. Such events occur naturally and all the time. The world has constantly been and is now splitting into different worlds each of which has been splitting, and so on. On Everett's terms, each of these worlds is as actual as your world and mine. If you think that your world is special because you're in it, think again. There are different worlds in which someone can claim identity and distinctiveness with as much Everettian justification as you do.

So, you may find the many worlds of Hugh Everett unimaginable. Not so Blake Crouch, who in his artful (though misnamed) novel *Dark Matter* has the physicist Jason Dessen discover a way of building "a twelve-foot cube the color of gunmetal" that can put human beings in superposition when they're in the box.[13] Jason is forced into the cube, and once in superposition, he can travel any of his life's roads not taken. Of those he chooses, some are frightening, others enchanting. The one he most desires to return to is the one he's been violently torn out of to be forcibly delivered into superposition.

OPENING THE DOOR TO RELATIVITY THEORY

Like Jason Dessen, all of us are imprisoned in a box of our own design that contains endless and unreal possibilities. It's located at the end of a development whose early stages Heidegger mentioned in 1949:

> All distances in space and time are shrinking. A place to which humans were previously underway for weeks and months, they now reach overnight on an airplane. What humans previously learned of after years or never at all, they now find out hourly and at once through the radio. The germinating and flourishing of plants that remained concealed through the seasons a film presents publicly and in one minute. Distant sites of the most ancient cultures are presented by a film as though they stood in the middle of today's traffic. Films moreover underscore their presenting by presenting the persons who are at work with the recording apparatus. The peak of the removal of all distance is reached by the television machinery that will soon pursue and dominate the entire structures and maneuvers of commerce.
>
> Humans reach what's farthest away in the shortest time. They cover the greatest distances and thus get everything at the shortest distance in front of themselves.[14]

What Heidegger did not foresee was the shrinking of distances to the distancelessness of a point in cyberspace. Today everything is neither distant nor near; everything is equally near and far on your computer or iPhone. You can call up the webcam on Turtle Bay on Oahu as easily and quickly as the webcam on the Belchen in the Black Forest. Adopting and adapting a distinction in geometry, we might say that cyberspace has a topology, but no metric. There are relations in cyberspace— ordering, sequence, nesting, and such, but there are no distances that have a metric, i.e., a rigid system of measurement.

Even in the world of tangible objects the shrinking of distances continues beyond what Heidegger could imagine. Airlines and interstate highways have further shortened travel times. Phone connections are quick. Ubiquity and instantaneity of resources are the goals of politicians, entrepreneurs, and engineers.

There are suggestive if slightly far-fetched analogies between Heidegger's philosophical reflections and Albert Einstein's scientific theory of Special Relativity. The German word *Ereignis* (event) is the one that Heidegger employed to name the rise of the modern era whose character Heidegger described as *Technik* (technology). And *Ereignis* is a term that kept haunting him in the margins of his first publication after World War II.[15] What for Heidegger designated the widest and deepest event of the modern era, showed up in Einstein's early German discourse as the minimal version of what an event could be, something that has no properties except the most austere identity, i.e., a point in the fourfold of space and time, a term that Einstein replaced with "point" and German *Punkt* once he had reached this country.[16] *Ereignis* as a mere point in the fourfold of spacetime is the conclusion of Heidegger's concerns with the shrinking of distances and is the very opposite of the rise and force of the epochal structure that occupied Heidegger in the mid-thirties of the twentieth century in fragments and essays of which he said: "The official title: Contributions to Philosophy and the essential Heading: Of the Event."[17]

STEPPING INTO THE WORLD OF SPECIAL RELATIVITY

Imagine a world that is empty except for two spaceships, one commanded by Alice, the other by you, you being Bob. If the two spaceships were to approach and touch each other, the imaginary world would consist of just one object with its one frame of reference, the Alice-Bob spaceship. As long as there are two objects, each with its own frame of reference, these two would have just one relation in common, their shared velocity.

If this is the world of Special Relativity, it would, to start with, have a property that would be strange to normal humans if they were to think about it, viz., that the world has an absolute speed limit. Naively we would think that velocities could be arbitrarily high. But in our world, the world of Special Relativity, nothing can move at a velocity higher than the speed of light. Though finite, it's a high speed indeed and to be intelligible to us we have to use a metric where the distance (over space) is very large or the time interval (in distance over time) is very small—186,000 miles over one second, or one foot over one nanosecond (one billionth of a second).

Still, we might get used to the idea of the speed of light. In fact, the idea that light signals may not be instantaneous has a venerable ancestry. In the West, it was the Greek philosopher Empedocles (c. 490–430 BCE) who first suggested that light has a velocity, and in the early eighteenth century there were already accurate computations of the speed of light. But there is another consequence of Special Relativity that entirely contradicts our spacetime intuitions. If you could follow Alice's clock, you would see that it runs slower than yours. This is not a feature of the timing device, rather time itself in Alice's frame of reference is slower.

The technical term for the slowing of time in Alice's spaceship is time dilation. The astronaut twin who spent a year in space showed measurably less aging than his earthbound twin. The twins were 254 miles apart, the distance of the Space Station from Earth, and the Space Station travels at 17,130 mph relative to Earth.

There is a thought experiment that yields a formula to compute the difference that the velocity between Alice and Bob, you being Bob, makes in the dilation of time in Alice's world; and in fact the formula lets you compute the (immeasurably small, it turns out) difference in aging between your aging and the aging of everything that's not you. The thought experiment has you imagine a photon clock. It's an imaginary device that consists of two parallel mirrors and a photon bouncing up and down between the mirrors.

You are invited to consider two settings. In the first the photon clock and you are in the same frame of reference. You (Bob) are to observe the photon clock with the photon moving from a point A at the bottom mirror to a point B at the top mirror back down to point A. Let's call the distance between the A and B *d*. This gets us equation [1]:

$$[1] \ \Delta t_b = \frac{2d}{c}$$

In the second setting you and the photon clock are in different frames of reference, and the photon clock is moving horizontally to the right from *E* to *G* at a velocity of *v*. From the point of view in your frame of reference, the photon is tracing the two equal sides of an isosceles triangle from a *E* to *F* to *G*. Let's call both the distance *E* to *F* and the distance *F* to *G L*. *L* can be stipulated within a range that can be defined by the extremes of the degrees of the angle that is enclosed by the two equal sides: When the degrees have reached zero, the two isosceles sides have merged into a vertical line. When the degrees have expanded to 180 degrees, the two isosceles sides have become a horizontal line.

Since (the length of) *L* is stipulated and *c* is known, t_a can be defined:

$$[2] \ \Delta t_a = \frac{2L}{c}$$

Call the base of the isosceles triangle, i.e., the distance from *E* to *G e*. That distance is equal to the product of the velocity *v* and the time interval t_a.

$$[3] \ e = v\Delta t_a$$

Half of the length of the base is:

$$[4] \ \frac{e}{2} = \frac{v\Delta t_a}{2}$$

where

$$\frac{v\Delta t_a}{2}$$

is the length of one of the legs of the right-angled triangle whose hypotenuse is L. This is merely a geometrical description. The Pythagorean theorem takes that description and provides an equation and entry into algebra:

$$[5] \quad L^2 = d^2 + \left(\frac{v\Delta t_a}{2}\right)^2$$

Now take the square root of (5):

$$[6] \quad L = \sqrt{d^2 + \left(\frac{v\Delta t_a}{2}\right)^2}$$

Combine (2) and (6) into (7) by replacing L in (2) by with the definition of L in (6):

$$[7] \quad \Delta t_a = \frac{2}{c}\sqrt{d^2 + \left(\frac{v\Delta t_a}{2}\right)^2}$$

Solve (1) for d:

$$[8] \quad d = \frac{c\Delta t_b}{2}$$

Square (8):

$$[9] \quad d^2 = \left(\frac{c\Delta t_b}{2}\right)^2$$

Combine equations (7), (8), and (9) by replacing d² in (7) with the definition of d² in (9):

$$[10] \quad \Delta t_a = \frac{2}{c}\sqrt{\left(\frac{c\Delta t_b}{2}\right)^2 + \left(\frac{v\Delta t_a}{2}\right)^2}$$

It's the linking of Alice's time interval, viz., Δt_a, with Bob's time interval, viz., Δt_b, through in equation (10) that results in gamma. Square equation (10) and get:

$$[11] \quad \Delta t_a^2 = \frac{4}{c^2}\left(\left(\frac{c\Delta t_b}{2}\right)^2 + \left(\frac{v\Delta t_a}{2}\right)^2\right)$$

Simplify (11), and the cancelations show:

$$[12] \quad \Delta t_a^2 = \Delta t_b^2 + v^2\left(\frac{\Delta t_a^2}{c^2}\right)$$

Solve (12) for $(\Delta t_b)^2$

$$[13] \quad \Delta t_b^2 = \Delta t_a^2 - v^2\left(\frac{\Delta t_a^2}{c^2}\right)$$

Extract Δt_a on the right side of (13):

$$[14] \quad \Delta t_b^2 = \Delta t_a^2 \left(1 - \frac{v^2}{c^2} \right)$$

Solve for $(\Delta t_a)^2$

$$[15] \quad \Delta t_a^2 = \frac{\Delta t_b^2}{\left(1 - \frac{v^2}{c^2} \right)}$$

Take the square root of (15):

$$[16] \quad \Delta t_a = \frac{\Delta t_b}{\sqrt{1 - \frac{v^2}{c^2}}}$$

Take Δt_b to the left side:

$$[17] \quad \frac{\Delta t_a}{\Delta t_b} = \frac{1}{\sqrt{1 - \frac{v^2}{c^2}}}$$

The right side of (17) is conventionally called gamma. Since in gamma the denominator will always be less than the numerator, Bob's time interval always has to be less than Alice's. Alice's is always larger than Bob's because Alice's time interval has been dilated.

Gamma also tells you that everything outside you is aging more slowly than you. Here is an illustration. How much more slowly does the world outside of you age if on average you move through your world at ten miles per hour?

$$[1] \quad v = 10 mph$$

Gamma gives the answer. First let's convert the ten mph in v to mps so that the expression of v conforms to the conventional way of rendering c.

$$[2] \quad v = 10/(60 \ minutes \times 60 \ seconds) = 0.002\ 777\ 777\ 8 \ mps$$

We limit the fraction to ten digits to the right of the decimal point. Next gamma requires us to square our v in mps:

$$[3] \quad v^2 = 0.000\ 007\ 716\ 0 \ mps^2$$

Next, we have to divide v² by c²:

$$[4] \quad \frac{v^2}{c^2} = 0.000\ 000\ 000\ 022\ 303\ 3$$

or roughly three quintillionths.

Next, gamma requires that we subtract v²/c² from one:

$$[5] \quad 1 - \frac{v^2}{c^2} = 0.999\ 999\ 999\ 97$$

The second to last step is to take the square root of the right side of (5):

$$[6] \quad \sqrt{1 - \frac{v^2}{c^2}} = 0.999\ 999\ 999\ 98$$

The last step is to divide one by the right side of the above equation:

$$[7] \quad 1/0.999\ 999\ 999\ 98 = 1.000\ 000\ 000\ 02$$

The moral of this story is that there is no discernible difference in aging between you and the rest of the world.

But what if v is 17,130 mph rather than ten mph as was the case for Scott Kelly who spent a year in the Space Station while his identical twin Mark stayed on Earth and when on Scott's return changes in Scott's genetic makeup were discovered?[18]

Let's do the math:

Again, gamma requires us to divide v² by c²:

$$[8] \quad \frac{v^2}{c^2} = 0.008\ 481\ 815\ 81$$

Then we have to subtract the right side of the above equation from one:

$$[9] \quad 1 - 0.008\ 481\ 815\ 81 = 0.991\ 518\ 184\ 191$$

Next, we need the square root of the right side of the above equation:

$$[10] \quad \sqrt{0.991\ 518\ 184\ 191} = 0.995\ 750\ 061\ 1$$

Finally, we have to divide one by the right side of the above equation:

$$[11] \quad 1/0.995\ 750\ 061\ 1 = 1.004\ 268\ 077\ 97$$

Thus, gamma tells us that:

$$[12] \quad \frac{(Scott\ (Alice))}{(Mark\ (Bob))} = 1.004\ 268\ 077\ 97$$

which means that a time interval for Scott has been longer than a time interval for Mark by four and a quarter thousandths, a difference that evidently showed up in Scott's genetic makeup.

Having stepped into the world of special relativity and traveled a ways across that world, we've come to the door of general relativity.

Editor's note: Albert Borgmann passed away with this afterword unfinished. Hopefully the parts that do exist help to demonstrate to a reader or educator that, with a basic algebra-level knowledge of math, we can step through the doors of quantum mechanics and relativity theory and gain some understanding of how they might apply to our world, thus helping to bridge the gap between physics and ethics and move us toward a moral cosmology (as illustrated in the conclusion).

NOTES

1. N. David Mermin, *It's About Time* (Princeton: Princeton University Press, 2005), p. 144.

2. Richard P. Feynman, *QED* (Princeton: Princeton University Press, 2006 [1985]), p. 14. See also p. 9.

3. Feynman, *QED*; R.I.G. Hughes, *The Structure and Interpretation of Quantum Mechanics* (Cambridge: Harvard University Press, 1992 [1989]); David Z Albert, *Quantum Mechanics and Experience* (Cambridge, Harvard University Press, 1992); Leonard Susskind and Art Friedman, *Quantum Mechanics: The Theoretical Minimum* (New York: Basic Books, 2014).

4. Einstein, "Zur Elektrodynamik bewegter Körper," *Annalen der Physik*, volume 17 (1905), pp. 891–921; "Die Feldgleichungen der Gravitation," *Sitzungsberichte der Preußischen Akademie der Wissenschaften*, 1915, pp. 844–47.

5. See Hughes for helpful pictures and accounts of the details on pp. 1–8.

6. Feynman, *QED*, p.9.

7. Frank Wilczek, *The Lightness of Being* (New York: Basic Books, 2008), p. 84.

8. Leonard Susskind and Art Friedman, *Quantum Mechanics: The Theoretical Minimum* (New York: Basic Books, 2014), p. 25.

9. Susskind, *Quantum Mechanics*, p. 52,

10. See Samuel Goudsmit: https://www.lorentz.leidenuniv.nl/history/spin/goudsmit.html.

11. See Hughes, pp. 5–7 and Susskind and Friedman, pp. 4–9.

12. Hugh Everett, III, "'Relative State' Formulation of Quantum Mechanics," *Review of Modern Physics*, volume 29, number 3 (July 1957), pp. 454–62.

13. Blake Crouch, *Dark Matter* (New York: Crown Publishers, 2016), pp. 121–26.

14. Heidegger, "Der Hinweis," in B*remer und Freiburger Vorträge* (Frankfurt: Klostermann, 1994 [1949]), p. 3. My translation.

15. See p. 178 in B*remer und Freiburger Vorträge*.

16. See Einstein, "Ist die Trägheit einer Körpers von seinem Energieinhalt abhängig?" *Annalen der Physik*, volume 18 (1905), p. 415; "Über das Relativitätsprinzip und die aus demselben gezogenen Folgerungen," *Jahrbuch der Radioaktivität und Elektronik,* volume 4 (1907), p. 455. And Heidegger, *Beiträge zur Philosophie (Vom Ereignis)* (Frankfurt: Klostermann, 1989).

17. Heidegger, *Contributions to Philosophy (Of the Event),* ed. Friedrich-Wilhelm von Herrmann (Frankfurt, Klostermann, 1989).

18. Francine E. Garrett-Bakelman et al. "The NASA Twins Study: A Multidimensional Analysis of a Year-long Human Spaceflight, *Science*, no. 364 (April 12, 2019), pp. 1–20.

Bibliography

Amrine, Frederick, et al. "Postscript: Goethe's Science: An Alternative to Modern Science or within It—or No Alternative at All?" *Goethe and the Sciences: A Reappraisal*, D. Riedel Publishing Company, Dordrecht, 1987.

Anderson, P. W. "More Is Different: Broken Symmetry and the Nature of the Hierarchical Structure of Science." *Science*, vol. 177, no. 4047, 1972, pp. 393–96, https://doi.org /10.1126/science.177.4047.393.

Aquinas, Thomas. *Summa Theologica*. Translated by Nicolaus Sylvius Bouillard and C.J. Drioux, 16th ed., Bloud and Barral, circa 1867–1874.

Aristotle. *On the Heavens*. Translated by W. K. C Guthrie, Harvard University Press, 2000.

———. *On the Soul*. Translated by Walter S. Hett, Harvard University Press, 1957.

———. *The Metaphysics*. Translated by Hugh Tredennick, Harvard University Press, 1958.

———. *The Nicomachean Ethics*. Translated by H. Rackham, Harvard University Press, 1957.

———. *The Physics*. Ed. & trans. by Philip H. Wicksteed and Francis M. Cornford, Harvard University Press, 1970.

Arthur, Gordon. "Religion and Values: Cosmic or Universal Ethics?" *Journal of Space Philosophy*, vol. 3, no. 2, 2014, pp. 23–30.

Audi, Robert, editor. *The Cambridge Dictionary of Philosophy*. Cambridge University Press, 1995.

"Beauty is Suffering [Part 1—The Mathematician]." YouTube, uploaded by astudyofevery thing, 27 September 2011, https://www.youtube.com/watch?v=i0UTeQfnzfM.

Behm, Wolf-Dieter Gudopp von. *Thales Und Die Folgen: Vom Werden Des Philoso- phischen Gedankens; Anaximander Und Anaximenes, Xenophanes, Parmenides Und Heraklit*. Königshausen Et Neumann, 2015.

Beiser, Frederick C. *Enlightenment, Revolution, and Romanticism*. Harvard University Press, 1992.

Bloomer, Kent C., et al. *Body, Memory, and Architecture*. Yale University Press, 1977.

Boeke, Kees. *Cosmic View: The Universe in 40 Jumps*. The John Day Company, 1957.

Borgmann, Albert, "The Headaches and Pleasures of General Education" (2003). Philosophy Faculty Publications. 12. https://scholarworks.umt.edu/philosophy_pubs/12.

———. *Holding on to Reality: The Nature of Information at the Turn of the Millenium*. The University of Chicago Press, 1999.

———. "Mind, Body, and World." *Philosophical Forum* vol. 8, 1976, pp. 68–86.

———. *Real American Ethics Taking Responsibility for Our Country*. University of Chicago Press, 2006.

Boyle, Nicholas. *Goethe: The Poet and the Age*. Vol. 2, Clarendon Press, 2000.

Boym, Svetlana. "Nostalgia and Its Discontents." *The Hedgehog Review*, vol. 9, no. 2, 2007, pp. 7–18.

Brentano, Clemens. "Der Philister." *Werke*, edited by Friedhelm Kemp, vol. 2, Hanser, Munich, 1963, pp. 959–1016.

Brittan, Gordon G. "Explanation and Reduction." *The Journal of Philosophy*, vol. 67, no. 13, 1970, p. 446., https://doi.org/10.2307/2023862.

Carnap, Rudolf. *Der Logische Aufbau Der Welt*. Felix Meiner Verlag, 1998.

Carroll, Sean. "Why Does the Universe Look the Way It Does?" *The Universe: Leading Scientists Explore the Origin, Mysteries, and Future of the Cosmos*, edited by John Brockman, Harper Perennial, New York, 2014.

Casey, Edward S. "The World of Nostalgia." *Man And World*, vol. 20, 1987, pp. 361–84.

Chaisson, Eric J. *Cosmic Evolution: The Rise of Complexity in Nature*. Harvard University Press, 2001.

Clark, Ella Elizabeth. *Indian Legends from the Northern Rockies*. University of Oklahoma Press, 1988.

Copernicus, Nicolaus. "On the Revolutions of the Heavenly Spheres." *Theories of the Universe*, edited by M. K. Munitz, The Free Press, 1957.

Crouch, Blake. *Dark Matter*. Crown Publishers, 2016.

Descartes, René. *Discourse on Method*. Translated by Laurence J. Lafleur, 2nd ed., The Library of Liberal Arts, 1956.

Donne, John. *The Anniversaries*. Frank Manley, editor. Baltimore: Johns Hopkins University Press, 1963.

Einstein, Albert. *The Collected Papers of Albert Einstein. Writings, 1918–1921*. Edited by Michel Janssen et al., Princeton University Press, 2002.

———. "Die Feldgleichungen Der Gravitation." *Sitzungsberichte Der Preussischen Akademie Der Wissenschaften*, Verl. d. Akademie d. Wiss, Berlin, 1915.

———. "Impact Lab." *Principle of Research, To the Physical Society, Berlin, for Max Planck's Sixtieth Birthday*, https://www.site.uottawa.ca/~yymao/impact/einstein.html.

———. "Ist Die Trägheit Eines Körpers Von Seinem Energieinhalt Abhängig?" *Annalen Der Physik*, vol. 323, no. 13, 1905, pp. 639–41, https://doi.org/10.1002/andp.19053231314.

———. "Über Das Relativitätsprinzip Und Die Aus Demselben Gezogenen Folgerungen." *Jahrbuch Der Radioaktivität Und Elektronik*, vol. 4, 1908.

———. "Zur Elektrodynamik Bewegter Körper." *Annalen Der Physik*, vol. 322, no. 10, 1905, pp. 891–921, https://doi.org/10.1002/andp.19053221004.

Engels, Friedrich, and Karl Marx. *Manifest Der Kommunistischen Partei*. Argument Verlag, 1999.

Esposito, Joseph L. *Schelling's Idealism and Philosophy of Nature*. Bucknell University Press Etc., 1977.

Everett, Caleb. *Numbers and the Making of US: Counting and the Course of Human Cultures*. Harvard University Press, 2017.

Everett, Hugh. "'Relative State' Formulation of Quantum Mechanics." *Reviews of Modern Physics*, vol. 29, no. 3, 1957, pp. 454–62, https://doi.org/10.1103/revmodphys.29.454.

Feynman, Richard P. *QED: The Strange Theory of Light and Matter*. Princeton University Press, 2006.

Fuentenebro de Diego, F., and Carmen Valiente Ots. "Nostalgia: A Conceptual History." *History of Psychiatry*, vol. 25, 2014, pp. 404–11.

Gadamer, Hans-Georg. *Truth and Method*. Ed. & trans. by Garrett Barden and John Cumming, The Crossroad Publishing Company, 1985.

Gadamer, Hans-Georg. *Wahrheit Und Methode*. Mohr Siebeck, 2010.

Galilei, Galileo. *Dialogue Concerning the Two Chief World Systems: Ptolemaic and Copernican*. Translated by Stillman Drake, 2nd ed., University of California Press, 1967.

———. "The Assayer." *The Controversy on the Comets of 1618: Galileo Galilei, Horatio Grassi, Mario Guiducci, Johann Kepler*, translated by C. D. O'Malley and Stillman Drake, University of Pennsylvania Press, Philadelphia, 1960.

Gardner, Martin, and Anthony Ravielli. *Relativity for the Million*. Pocket Books, 1965.

Goethe, Johann Wolfgang von. *Die Wahlverwandtschaften*. Vol. 6, Christian Wegner, 1968.

———. *Faust*. Christian Wegner, 1966.

———. *Goethes Werke: Hamburger Ausgabe*. Christian Wegner, 1966.

———. *Wilhelm Meisters Wanderjahre*. Beck, 1973.

———. *Von Der Weltseele Eine Hypothese Der Höhern Physik Zur Erklärung Des Allgemeinen Organismus (1798)*. 2nd ed., Friedrich Perthes, 1806.

———. "Ideen Zu Einer Philosophie Der Natur." *Schellings Werke*, edited by Schröter Manfred, Beck, München, 1956.

Goethe, Johann Wolfgang von,, and Johann Peter Eckermann. *Goethes Gespräche Mit Eckermann*. Edited by Franz Deibel, Insel-Verlag, 1949.

Goodman, Nelson. *The Structure of Appearance*. 1970.

Goudsmit, Samuel A. "The Discovery of the Electron Spin." *Foundations of Modern EPR*, 1998, pp. 1–12, https://doi.org/10.1142/9789812816764_0001.

Gould, Stephen Jay. *Rocks of Ages: Science and Religion in the Fullness of Life*. Ballantine Books, 2011.

Greene, Brian. *The Elegant Universe: Superstrings, Hidden Dimensions, and the Quest for the Ultimate Theory*. Norton, 1999.

———. *The Elegant Universe*. Norton, 1999.

———. *The Fabric of the Cosmos: Space, Time, and the Texture of Reality*. Alfred A. Knopf, 2004.

Grinnell, George Bird. *Blackfeet Indian Stories*. Riverbend, 2005.

Güdel, Helen. *Lieber Alex: Briefe aus dem Walliser Bergdorf Törbel*. Vol. 2, Atlantis Verlag, 1993.

———. *Lieber Alex: Briefe aus dem Walliser Bergdorf Törbel*. Vol. 3, Atlantis Verlag, 1995.

———. *Lieber Alex: Von Menschen Und Tieren Im Walliser Bergdorf törbel*. Vol. 1, Atlantis Verlag, 1991.

Hawkings, Steven. *A Brief History of Time*. Bantam Books, 1988.

Heidegger, Martin. *Anmerkungen I-V (Schwarze Hefte 1942–1948)*. Edited by Peter Trawny, V. Klostermann, 2015.

———. *Beiträge Zur Philosophie (Vom Ereignis)*. V. Klostermann, 1989.

———. *Bremer Und Freiburger Vorträge*. Edited by Petra Jaeger, V. Klostermann, 1994.

———. *Contributions to Philosophy (of the Event)*. Edited by Friedrich-Wilhelm Von Herrmann, V. Klostermann, 1989.

————. *Die Frage Nach Dem Ding: Zu Kants Lehre Von Den Transzendentalen Grundsätzen.* Max Niemeyer, 1962.

————. *Feldweg-Gespräche: 1944/45.* Edited by Ingrid Schüssler, V. Klostermann, 1995.

————. *Identität Und Differenz.* Neske, 1957.

————. "Die Frage Nach Der Technik." *Vortrage Und Aufsatze,* Neske, Pfullingen, 1954.

Hofer, Johannes. *Dissertatio Medica De Nostalgia, oder Heimwehe.* Jacob Bertschi, 1688.

Hughes, R. I. G. *The Structure and Interpretation of Quantum Mechanics.* Harvard University Press, 1992.

Hume, David. *A Treatise of Human Nature.* Clarendon Press, 1968.

Kant, Immanuel. *Ausgewählte Kleine Schriften.* Edited by Horst D. Brandt, Felix Meiner, 1999.

————. *Grundlegung Der Metaphysik Der Sitten. Akademie-Ausgabe.* Georg Reimer, 1911.

————. *Kritik Der Praktischen Vernunft Text Der Ausgabe Von 1787.* Edited by Karl Vorländer, Felix Meiner, 1963.

————. *Kritik Der Reinen Vernunft Text Der Ausgabe Von 1781.* Edited by Raymund Schmidt, Felix Meiner, 1956.

————. *Metaphysische Anfangsgründe der Naturwissenschaft,* Hartknoch, 1787.

————. *Universal Natural History and Theory of the Heavens,* in *Theories of the Universe: From Babylonian Myth to Modern Science,* Milton K. Munitz, ed. New York: The Free Press, 1957, p. 237.

————. *Universal Natural History and Theory of the Heavens, or an Attempt to Account for the Constitutional and Mechanical Origin of the Universe upon Newtonian Principles* (1755), in *Kant: Natural Science,* Eric Watkins, ed. Cambridge University Press, 2012.

————. *Metaphysical Foundations of Natural Science* (1786). Cambridge University Press, 2004, translated and edited by Michael Friedman.

————. *Was Ist Aufklärung?* Felix Meiner, 1999.

Kaur, Tejinder, et al. "Teaching Einsteinian Physics at Schools: Part 1, Models and Analogies for Relativity." *Physics Education,* vol. 52, no. 6, 2017, p. 065012, https://doi.org/10.1088/1361–6552/aa83e4.

Kellert, Stephen H. *In the Wake of Chaos: Unpredictable Order in Dynamical Systems.* University of Chicago Press, 1993.

Kepler, Johannes. *Astronomia Nova.* Translated by William H. Donahue, Green Lion Press, 2015.

————. "Praefatio Antiqua Ad Lectorem." *Mysterium Cosmographicum,* Dalton House, Paris, 2018.

Lakoff, George, and Mark Johnson. *Metaphors We Live By.* University of Chicago Press, 1980.

Lassiter, Cisco. "Relocation and Illness: The Plight of the Navajo." *Pathologies of the Modern Self: Postmodern Studies on Narcissism, Schizophrenia, and Depression,* edited by David Michael Levin, New York University Press, New York, 1987.

Lewis, Meriwether, and William Clark. *The History of the Lewis and Clark Expedition.* Edited by Elliott Coues, vol. 2, Dover Publications, 1998.

Lovejoy, Arthur O. *The Great Chain of Being: A Study of the History of an Idea.* Harvard University Press, 1964.

Maritain, Jacques. *Moral Philosophy.* Geoffrey Bes, 1956.

McMullin, Ernan. "Cosmology." *Routledge Encyclopedia of Philosophy,* edited by Edward Craig, Routledge, 1998, pp. 677–81.

Mermin, N. David. *It's about Time: Understanding Einstein's Relativity.* Princeton University Press, 2005.

Moore, Kathleen Dean. *Great Tide Rising: Towards Clarity & Moral Courage in a Time of Climate Change.* Counterpoint, 2016.

Morrison, Philip, and Phylis Morrison. *Powers of Ten: About the Relative Size of Things in the Universe.* Scientific American Library, 1982.

Nabokov, Peter. *Restoring a Presence: American Indians and Yellowstone National Park.* University of Oklahoma Press, 2004.

Nagel, Ernest. *The Structure of Science.* Harcourt, Brace & World, 1961.

Netting, Robert MacCorkle. *Balancing on an Alp.* Cambridge University Press, 1981.

Newton, Isaac. *Newton: Philosophical Writings.* Edited by Andrew Janiak, Cambridge Univ Press, 2004.

———. *Opticks.* Prometheus Books, 2003.

———. *The Principia: Mathematical Principles of Natural Philosophy.* Translated by Bernard Cohen and Anne Whitman, University of California Press, 1999.

Pack, Robert. *Before It Vanishes: A Packet for Professor Pagels.* Godine, 1989.

Pagels, Heinz R. *The Cosmic Code: Quantum Physics as the Language of Nature.* Bantam, 1982.

Panek, Richard. "How a Dispute over a Single Number Became a Cosmological Crisis." *Scientific American*, Mar. 1, 2020, pp. 30–37.

Pascal, Blaise. "Le Mémorial." *Pensées*, Edited by Victor Giraud, Rombaldi, 1935.

Primack, Joel R., and Nancy Ellen Abrams. *The View from the Center of the Universe: Discovering Our Extraordinary Place in the Cosmos.* Riverhead Books, 2006.

Richards, Robert J. "Nature Is the Poetry of the Mind, or How Schelling Solved Goethe's Kantian Problems." *The Kantian Legacy in Nineteenth-Century Science*, edited by Michael Friedman and Alfred Nordmann, MIT Press, 2006.

Rosenberg, Alexander. *The Atheist's Guide to Reality: Enjoying Life Without Illusions.* W.W. Norton, 2011.

Ryle, Gilbert. *The Concept of Mind.* Barnes & Noble, 1968.

Sandkühler, Hans Jörg. *Friedrich Wilhelm Joseph Schelling.* Metzler, 1970.

Schelling, Friedrich Wilhelm Joseph, *Ideas for a Philosophy of Nature (Texts in German Philosophy)*, Errol E. Harris and Peter Hearth, translators, 2nd edition, Cambridge University Press, 1988.

———. "Fernere Darstellungen aus dem System der Philosophie." *Ausgewählte Schriften*, edited by Manfred Frank, vol. 2, Suhrkamp, Frankfurt, 1985.

———. "Ueber Den Wahren Begriff Der Naturphilosophie." *Ausgewählte Schriften*, edited by Manfred Frank, vol. 2, Suhrkamp, Frankfurt, 1985.

Sellars, Wilfrid. "Philosophy and the Scientific Image of Man." *Frontiers of Science and Philosophy*, edited by Robert Garland Colodny, University of Pittsburgh Press, Pittsburgh, 1962, pp. 35–78.

Shute, Nevil. *On the Beach.* Heinemann, 1957.

Skinner, Charles M. *Myths and Legends of Our Own Land.* J.P. Lippincott Co, 1896.

Smeenk, Christopher, and George Ellis. "Philosophy of Cosmology." *Stanford Encyclopedia of Philosophy*, Stanford University, 26 Sept. 2017, https://plato.stanford.edu/entries/cosmology/.

Smith, Kelly C. "Cosmic Ethics: a Philosophical Primer." *Workshop Report: Philosophical, Ethical, and Theological Implications of Astrobiology*, edited by Nancy Roth et al., American Association for the Advancement of Science, 2007.

Sunderlee, Charles R. "A Thrilling Event on the Yellowstone." *The Helena Daily Herald*, 18 May 1870.

Susskind, Leonard, and George Hrabovsky. *The Theoretical Minimum: What You Need to Know to Start Doing Physics.* Basic Books, 2013.

Swimme, Brian, and Thomas Berry. *The Universe Story.* HarperSanFrancisco, 1994.

Taylor, Charles. *A Secular Age.* Harvard University Press, 2007.

Tegmark, Max. "Parallel Universes." *Scientific American,* May 2003, pp. 40–51.

Thales. "Thales." *Die Fragmente Der Vorsokratiker,* ed. & trans. by Hermann Diels, vol. 1, Weidmann, Zürich, 1964, pp. 67–81.

Theil, Stefan. "Trouble in Mind." *Scientific American,* vol. 313, no. 4, 2015, pp. 36–42, https://doi.org/10.1038/scientificamerican1015-36.

Thomas, William Cave. *Ethics or the Mathematical Theory of Evolution Showing the Full Importance of the Doctrine of the Mean, and Containing the Principia of the Science of Proportion.* Smith, Elder, & Co., 1896.

Tipler, Frank J. *The Physics of Immortality.* Anchor Books, 1995.

Tucker, Mary Evelyn and Kathleen Dean Moore. "A Roaring Force from One Unknowable Moment." *Orion Magazine,* no. 3, May 2015.

Veblen, Thorstein. *The Theory of the Leisure Class.* Transaction Publishers, 2000.

Veisdal, Joergen. "Einstein and Hilbert's Race to Generalize Relativity." *Medium,* Medium, 11 Sept. 2022, https://jorgenveisdal.medium.com/einstein-and-hilberts-race-to-generalize-relativity-6885f44e3cbe.

Virgil. *Aeneid.* Edited by G.P. Goold. Translated by H. Rushton Fairclough, Harvard University Press, 1994.

von Weizsäcker, Carl Friedrich. "Nachwort." *Die Einheit Der Natur Studien,* Christian Wegner, Hamburg, 1971.

Weber, Max. *Wissenschaft Als Beruf.* Reclam, 1995.

Weinberg, Steven. *Dreams of a Final Theory.* Pantheon Books, 1992.

———. *First Three Minutes: A Modern View of the Origin of the Universe.* 2nd ed., Basic Books, 1993.

———. *To Explain the World.* Harper Collins, 2015.

———. "A Model of Leptons." *Physical Review Letters,* vol. 19, no. 21, 20 Nov. 1967, pp. 1264–66, https://doi.org/10.1103/physrevlett.19.1264.

Welch, James. *Fools Crow.* Viking, 1986.

Whittlesey, Lee H. "Native Americans, the Earliest Interpreters: What Is Known About Their Legends and Stories of Yellowstone Park and the Complexities of Interpreting Them." *The George Wright Forum,* vol. 19, no. 3, 2002.

Wigner, Eugene P. "The Unreasonable Effectiveness of Mathematics in the Natural Sciences." *Communications on Pure and Applied Mathematics,* vol. 13, no. 1, 1960, pp. 1–14, https://doi.org/10.1002/cpa.3160130102.

Wilczek, Frank. *The Lightness of Being: Mass, Ether, and the Unification of Forces.* Basic Books, 2008.

Wilson, Edward O. *Sociobiology: The New Synthesis.* Belknap Press of Harvard University Press, 2000.

Xin, Xianhao. Glashow-weinberg-salam model: An example of electroweak symmetry breaking. 2007, https://guava.physics.uiuc.edu/~nigel/courses/569/Essays_Fall2007/files/xianhao_xin.pdf

Index

Page numbers in *italics* refer to tables.

About the Author

Albert Borgmann was born in Freiburg, Germany, in 1937. He writes that he was raised "in the shadow of the Gothic cathedral, of the Black Forest, and of the university where Husserl and Heidegger had been teaching." Borgmann began his college career at the University of Freiburg, attending Heidegger's lectures in 1957, but later transferred to the University of Texas. Borgmann wanted to see more of the United States and went on to earn an MA in German literature from the University of Illinois, Urbana, in 1961, where he met his future wife, Nancy Borgmann (nee Quasthoff). After marrying in Freiburg, Albert and Nancy moved to Munich, Germany, where Albert earned a PhD in philosophy from the University of Munich. They then returned to the United States. Borgmann briefly taught German literature at the University of Illinois, and then philosophy at DePaul University and the University of Hawai'i. In 1970, he settled with his family in Missoula, Montana. There he taught philosophy to a generation of students at the University of Montana, publishing widely, and becoming Regents Professor in 1996, the third person to receive the honor in the university's history. Albert's beloved wife Nancy passed away in 2009. Albert retired from teaching philosophy in 2020, after fifty years at the University of Montana.

Borgmann's work concerns the philosophy of society and culture, with particular emphasis on the cultural force of technology. His 1984 book, *Technology and the Character of Contemporary Life*, has become a landmark text, not only in the philosophical study of technology but also for a wide range of the humanities and social sciences. His other major publications include *Crossing the Postmodern Divide* (1992), *Holding on to Reality: the Nature of Information at the Turn of the Millennium* (1999), *Power Failure* (2003), and *Real American Ethics* (2006).

Aside from teaching and publishing, Borgmann was an influential force in state and local politics as well as administration at the University of Montana. Borgmann lived for over five decades on a mountain near the Rattlesnake Creek in Missoula, always in the company of a couple of large dogs.

Printed in the USA
CPSIA information can be obtained
at www.ICGtesting.com
LVHW050307250124
769324LV00003BA/8

9 781666 900460